The Institution of
StructuralEngineers

Car park design

AF874498

Authors

P Guest BSC MSc DipTE
D Gullock BA (Hons) M.Arch PG Cert ARB (Fatkin Ltd)
A McBryan BEng (Hons) CEng MIET (AECOM)
F Petch BA (Hons) M.Arch PG Cert ARB (Fatkin Ltd)
M Pundsack BEng (Hons) CEng MRICS AIFireE FIStructE (City of London) *Chair*
P Robery BSc PhD FREng CEng FICE FICT FCS (Robery Forensic Engineering Ltd)
R Simmons BSc (Hons) MCIOB CBuildE MCABE FCIHT FBPA AMICE (Ballast Nedam UK)
J Stewart CEng MIStructE (Ballast Nedam Parking Ltd)
C Watkins BA (Hons) MA NRAC (Consultant) (Arup)
C Whapples BSc (Hons) CEng FIStructE FICE (Stripe Consulting) *Overseeing Consultant*

Reviewers

N Ely BSc (Hons) CEng MIStructE MCIOB (Environment Agency)
G Deuce BEng CEng FIStructE FICE (Mace Group)

Publishing

L Baldwin BA (Hons) DipPub (The Institution of Structural Engineers)

Published by The Institution of Structural Engineers
International HQ, 47–58 Bastwick Street, London EC1V 3PS, United Kingdom
T: +44(0)20 7235 4535
E: mail@istructe.org
W: www.istructe.org

First published (version 1.0) June 2023
978-1-906335-62-5 (print)
978-1-906335-63-2 (pdf)

The authors

Mark Pundsack – Chair
Mark is Assistant District Surveyor at the City of London. He is an experienced Building Control Officer with a particular interest and specialism in structural engineering and fire. He has been involved in assessing some of the largest and most complex schemes in the country, including a number of very tall commercial towers. This involvement has spread from early design, through approvals, inspections from foundations to occupation, and some interesting factory inspections in Europe and China. He manages a group of multidisciplinary building control professionals; ensuring the safety, accessibility and energy efficiency of buildings withing the City of London.

He is vice chair of the IStructE's Safety, Health and Wellbeing Panel, and a member of the Business Practice and Regulatory Control Committee. Since the Hackitt report, Mark has been much involved in providing advice to government on the future of building control, including being a member of Raising The Bar WG6, and the Future of Building Control Working Group.

Chris Whapples – Overseeing Consultant
Chris has more than 50 years' experience as an independent Parking Consultant and was involved in the preparation of the 2nd, 3rd and 4th editions of the IStructE's *Design recommendations for multi-storey and underground car parks* guidance.

He has been responsible for the design and construction of many car parks both in the UK and internationally, and was deeply involved in the development of the patented Vertical Circulation Module (VCM) system. He was a member of three winning teams for the Best New Car Park Award and the Best Refurbishment Award before progressing to the British Parking Award's panel of judges.

Representing the British Parking Association (BPA) Chris worked in association with the Transport Research Laboratory in developing new standards for testing vehicle impact barriers, and with the Building Research Establishment in evaluating fire spread in car parks.

Specialist knowledge of car parks has enabled him to co-author the ICE's *Recommendations for the Inspection, Maintenance and Management of Car Park Structures* (2nd edition), and to sit on several European Technical Committees relating to durability, repair, inspection and maintenance of concrete structures. He is a joint author of the first edition of the *Car Park Designers' Handbook*, and has written and presented several specialist papers to promote the betterment of car parks.

Chris was chairman of the BPA's Parking Structures Group. He was presented the Ernest Davies Award in 2006 for advancing parking knowledge, and ultimately the BPA's Lifetime Achievement Award in 2015, for his outstanding contribution to the sector.

The authors

Peter Guest
Peter has been involved in parking for more than 50 years. Starting in policy research at the GLC in 1972 he was later involved in street parking management. He became involved in parking technology – as a member of, and later chairman of – the British Standards Committee, and the UK's representative on the European committee. He moved to consultancy in 1987 and worked both in the UK and internationally, including World Bank studies in Europe and Asia. He joined a specialist car park design consultancy in 1997 as a partner, working on projects dealing with all aspects of parking. Since 2003 Peter has been self-employed. During this period, he designed and procured the Abu Dhabi Parking programme MAWAQiF, probably the biggest parking project in the world. In 2023 Peter received the BPA's Lifetime Achievement Award. He is now semi-retired but continues to write for *Parking Today*, and is a member of the Temecula Group; a US parking industry think tank.

Dan Gullock
As project architect at Arup Associates and Hawkins\Brown, Dan worked on nationally significant transport infrastructure projects including the High Speed 2 Birmingham Interchange and Liverpool Street and Tottenham Court Road Crossrail Stations. As co-director and founder of Fatkin, he has subsequently designed and delivered multiple transport hub and movement strategy projects across the UK for clients including Network Rail, various NHS Trusts, Local Authorities, universities and major developers. Dan regularly undertakes architectural CRE duties on Network Rail projects and has an extensive knowledge of car park construction. He is passionate about innovation and design quality.

Alan McBryan
Alan is a Regional Director in the AECOM London office with over 25 years' experience in the design and delivery of building services projects. With a broad range of experience from master-planning and concept designs through to detail design and construction activities, Alan's experience on projects ranges from offices, data centre/mission critical, mixed use, retail, through to industrial and leisure projects and of course car parks. An electrical engineering graduate from The University of Birmingham, Alan has spent part of his career in Dublin and Singapore, but is most familiar with the UK, having joined the AECOM London team in 2004. Alan's recent experience is in leading AECOM teams on major projects and he is comfortable dealing with all aspects of buildings including sustainability and fire, as well as core building services systems.

Fiona Petch
Fiona is a co-director and founder of Fatkin, an architecture practice specialising in transport infrastructure and healthcare design. She is experienced in planning strategy, stakeholder engagement and public consultation. Based in London, Fiona is passionate about active travel and sustainable design in transport across rural and urban areas. She is committed to broadening access to architecture and transport design for people not traditionally represented in these industries, and is a mentor to young people and visiting lecturer at institutions including the London School of Architecture and the University of the West of England.

Peter Robery
Peter is currently founder director of Robery Forensic Engineering Ltd and Visiting Professor in Forensic Engineering at the University of Birmingham. Formerly he was Technology Director for Transportation-Europe with Halcrow/CH2M Hill specialising in concrete materials technology, durability, inspection, testing, repair and protection of structures. He has worked for Maunsell/AECOM, Harry Stanger and Taylor Woodrow Construction. A Past President of The Concrete Society, Peter is currently Chair of BSI B/517/8 (Concrete protection and repair). He teaches concrete deterioration, maintenance and repair at the Universities of Birmingham and Leeds. Peter served on the Task Group of the IStructE's 3rd and 4th editions of *Design recommendations for multi-storey and underground car parks*, as well as the Steering Committee for the first and second editions of the ICE's *Recommendations for the Inspection, Maintenance and Management of Car Park Structures*.

The authors

Russell Simmons
A Chartered Construction Manager and Chartered Building Engineer, Russell is the founder of Stripe Consulting who specialise in the car park built environment. He has been chair of the Parking Structures Group at the BPA for 10 years and also chairs the suicide prevention group and fire risk in car parks group. He has contributed to industry publications, appeared on television and spoken at many high-profile events on the subject of car park design and construction. He is a winner of Parking Professional of the Year, and Young Parking Professional awards. In 2019 Russell joined the large construction group Ballast Nedam and started Ballast Nedam Parking Ltd in the UK and grew the car park design and build division from £0 to over £20M revenue. Russell is currently CEO of Ballast Nedam UK, which is the parent group for multiple Ballast Nedam business units in the UK.

Jeff Stewart
A Chartered Structural Engineer and past chair of the IStructE's Southern Regional Group, Jeff is currently the Technical Director of Ballast Nedam Parking Ltd, part of the Ballast Nedam UK Group, who specialise in the design and construction of multi-storey car parks. He has over 25 years' experience in the design and construction of multistorey car parks — having held senior roles in some of the UK's leading car park specialist design-and-build contractors over this period. Jeff was also part of the IStructE Task Group responsible for the 4th edition of *Design recommendations for multi-storey and underground car parks*.

Chris Watkins
Chris is a senior consultant within Arup's Access and Inclusive Design team. He has been working in the equality and inclusion field with a focus on disability equality since 2010, including as an NRAC accredited access consultant since 2018. He currently specialises in inclusive design in public realm and transport infrastructure, focusing on streetscape, rail and aviation sectors. His current projects include assessing the impact of public infrastructure on equality and inclusion for different gender and LGBTQIA+ groups, and ensuring that the needs of people who rely on private vehicles and cycles as a mobility aid are considered in the design of our towns and cities.

Acknowlegements

The Institution would like to thank Matthew Shum for his work in creating:
Figures 4.4, 5.1, 5.4–5.11, 11.1, 11.2, 11.4, 11.5, 11.7–11.10 and 11.14.

Permission to reproduce the following has been obtained, courtesy of these individuals/organisations:
Front cover photo: Veranda car park, Rotterdam, The Netherlands ©Paul de Ruiter Architects, photography by Rien van Rijthoven

Figure 0.1: Historic England Archive
Figure 0.2: ©Health and Safety Executive. Contains public sector information licensed under the Open Government Licence v3.0.
Figures 1.1, 1.3b, 5.13–5.24, 9.1, 12.3, 12.6 and 16.1: ©Stripe Consulting
Figure 1.4b: ©Aliaksandr Yarmashchuk (iStock)
Figure 1.5: ©Beeldbewerking (iStock)
Figure 4.1: ©bilanol (iStock)
Figure 4.2: ©Trainman111 (iStock)
Figure 4.3: ©Richard Chivers Photography
Figure 4.5: ©rumruay (Adobe Stock)
Figure 4.7: ©Andrii Shepeliev (iStock)
Figure 4.8: ©MaHa Heang 245789 (Adobe Stock)
Figure 4.10: ©Triflex (UK) Ltd
Figure 5.2: ©Andrii Koval (iStock)
Figure 5.3a: Cycle-works
Figure 5.3b: ©y-studio (iStock)
Figure 7.1: ©Merseyside Fire & Rescue Service
Figure 8.1: Derived/adapted from PAS 2080 (BSI)
Figures 8.2 and 8.3: ©Greater London Authority. Contains public sector information licensed under the Open Government Licence v3.0.
Figure 8.4: ©Depict Photography
Figure 8.5: ©Rasmus Hjortshøj (Coast Studio)
Figure 9.2: ©British Parking Association
Figure 9.3: ©silkwayrain (iStock)
Figure 11.3: Paul Bunch/Alamy Stock Photo
Figure 11.6: Bourne Group Ltd
Figure 11.11: ©cbglp2 (iStock)
Figure 11.12: ©Casanowe (iStock)
Figure 12.5: ©Ballast Nedam Parking Ltd
Figures 13.1a, 13.1b and 13.2: Derived/adapted from BS 7346-7 (BSI)
Figure 13.3: ©Fahroni (iStock)
Figure 16.2: ©INTEGRA-pw/projekt w - Systeme aus Stahl GmbH
Figure 17.1: ©Sika
Figure 17.2: ©Concrete Repairs Ltd
Figure 18.1: ©Accuris
Figure 20.1a: Wikimapia.org
Figures 20.2, 20.3 and 20.5: Wohr Parking Systems

Table 1.1: ©Rushmoor Borough Council
Table 3.1: Derived/adapted from Table 2 of the Department for Transport's *Traffic Advisory Leaflet 5/95*.
Table 8.1: Eco.Zero™ (AECOM)
Table 11.3: Derived/adapted from BS 8500 (BSI)
Table 11.5: Derived/adapted from the NSCS and Table F4 of BS 13670 (BSI)
Table 17.1: ©The Concrete Society (TR22 4th ed, 2010).

Permission to reproduce extracts from British Standards is granted by BSI. British Standards can be obtained in PDF or hard copy formats from the BSI online shop: www.bsigroup.com/Shop

Contents

Foreword

It has been my pleasure to chair the contributors group for this *Car park design* guidance. As a Building Control Officer (and therefore not a car park expert!) it has been a huge learning experience and I marvel at the depth and breadth of knowledge that The Institution of Structural Engineers can convene.

The brilliant and expert contributors group have produced something which is useful and meaningful to everyone who is involved in the design of car parks.

We have deliberately split the book into 'conceptual' and 'detailed' design so that those who wish to understand the concepts, such as clients, project managers and architects can do so without having to wade through the nitty gritty. We also ensure there is enough nitty gritty for structural engineers, services engineers and those involved in the delivery of projects to get their teeth into.

It is also my fervent wish that students and those starting out in their careers will find this book as illuminating as I have found the process of putting it together to be.

Mark Pundsack
Chair

Foreword

It is my pleasure and privilege to introduce this *Car park design* guidance.

I have known some of the contributors for several years and they have impressed me with their ability and experience in the world of parking structures. They are leading contributors to the British Parking Association's Parking Structures Group.

Being a civil engineer, I admire the tenacity of my professional colleagues and their peers to design and deliver ever safer parking structures. As a motorist, I simply want to be able to park my car efficiently and effortlessly somewhere I feel intuitively safe and welcome. I want to be able to get to and from my parking spot within the car park easily. Also, I want car parks (especially those contained within buildings) to improve the public realm and not be an eyesore.

It is fitting that The Institution of Structural Engineers has chosen to completely revise, refresh and rename its original publication *Design recommendations for multi-storey and underground car parks.* So much has changed since it was originally published. Simply updating and creating a fifth edition would be to ignore contemporary consumer expectations and the new materials now being used in the construction of vehicles and buildings.

There are important safety and economic implications of the increasing size and weight of the "average vehicle" up from 1.5t in 1974 to almost 2t today. These increased loadings and the changing fire safety requirements are all considerations not just for new car parks but in the thousands which already exist today.

Environmental sustainability means energy-saving lighting, water recycling and air quality monitoring becoming desirable features. Add in greenery from planters and pots through to living walls, and the look and feel of our car parks is transforming. We need to support car sharers and active travel choices with special parking bays, including secure cycle and motorcycle parking and EV charging.

Accessibility and equalities considerations mean special bays for people with disabilities, parents/carers with children, better way-marked pedestrian routes and more.

Crucially, the health and welfare of everyone who uses our parking facilities is paramount, and so out of necessity, sadly, there is also a need to include advice and guidance on managing the risks of people falling from height and the prevention of suicide in car parks.

This guide will be a valuable resource and reference for car park architects, engineers, developers, funders, and those who service and maintain these facilities. It should be of interest to parking operators in the public and private sector, across central and local government. Likewise in retail and leisure; healthcare and higher education; railways and airports — in fact, everyone who needs to understand what is going on in the world of car park design.

To some, it's a matter of creating something which is safe and sound, to others it is a work of art: in combination we have the potential for parking masterpieces.

There seems to be everything you need to know in this publication, which is a world leader in its advice, advocacy, and approach and which should be the 'go to' reference for everyone with an interest in new and existing car parks. I commend it to you all.

Kelvin Reynolds

Kelvin Reynolds, I.Eng., FIHE, FRSA, FBPA, DipHTE (Middx).
Chief Technical Services and Governance Officer
British Parking Association

british parking awards 2016 WINNER Lifetime Achievement Award.

Definitions and acronyms

Accessway: carriageway not adjoining bays and used solely for the movement of vehicles.

Aisle: an accessway serving adjoining bays.

Angled parking: where parking bays are at an angle less than 90° to drive aisle.

ANPR: automatic number plate recognition.

Bay: parking area, exclusive of aisle or other adjoining area, allocated to one vehicle.

Bay turnover: number of times a parking bay is occupied within a 24-hour period.

Benchmark inspection: post-construction assessment of a car park's durability — entailing the comparison of as-built structural records and survey results, with the original designs and specifications in the handover information pack (HIP) — enabling future inspection and maintenance programmes to be defined.

Bin: two rows of bays with the access aisle running between them. A half-bin is one row of bays and the aisle serving them.

CE mark: a form of product identification that:

- Indicates that the manufacturer has checked the product meets EU safety, health or environmental requirements
- Is an indicator of a product's compliance with EU legislation
- Allows the free movement of products within the European market

Note: from 1 January 2025 the CE mark is expected to be replaced by the UK conformity assessed (UKCA) mark, in England, Scotland and Wales.

Clearway ramp: a ramp system that does not include an aisle in its circulation and which provides unencumbered access between the parking floors and an entrance or exit.

Churn: see **Bay turnover**.

Condition survey: visual examination of all accessible areas and associated fabric/elements, and measurement of evidence of deterioration, including any structural investigation.

Cross aisles: aisles which interconnect the main drive aisles (generally at 90°).

Daily surveillance: visual observation and reporting of equipment breakdown, obvious deterioration and damage to the structure, cladding and edge protection, and vandalism/general misuse of the facility.

Deck: Slab or floor at any level.

Design service life: intended working life of the structure, as set out in the design brief, on which recommended values for parameters such as concrete class and minimum cover for durability have been chosen to give a low-risk of reinforcement becoming excessively corroded and requiring significant repair.

Dynamic capacity: applicable to individual parts of a car park or to the building as a whole. It is the maximum flow per hour of vehicles (or where appropriate, people) that can be accommodated.

Dynamic efficiency: measure of the ability of a car park to process vehicles under standard operating conditions.

Echelon parking: see **Angled parking**.

Gross vehicle mass/weight: mass/weight of vehicle plus passengers and luggage.

Ground clearance: shortest distance between lower part of vehicle body/chassis and road when vehicle is unladen.

Inclusive design: the design of mainstream products and/or services that are accessible to, and usable by, as many people as reasonably possible without the need for special adaptation or specialised design.

Inspector: person competent by virtue of a combination of qualification, experience and training to undertake **condition surveys**. May be employed by the same or a different organisation from the engineer. Will usually be an experienced (Chartered) civil or structural engineer.

Lean design: avoiding overdesign and achieving greater utilisation ratios of structural elements, including foundations.

Life-care plan: long-term plan for the implementation of an inspection maintenance strategy.

Lighting efficacy: measure of how well a light source produces visible light.

Local planning authority (LPA): local government body empowered by law to exercise urban planning functions for a particular area.

MSCP: multi-storey car park.

Oblique parking: see **Angled parking**.

PARCS: parking access and revenue control systems.

Parking angle: angle between the length of a bay and the aisle from which it is served.

Parking ramp: inclined section of deck that forms part of a ramping system and has a gradient shallower than 1:20.

Pedestrian ramp: inclined surface for use by pedestrians to access a change in level.

Ramp: accessway (or less commonly **aisle**) connecting parking areas at different levels.

Reservoir: accessway where cars may queue without obstructing movements in other parts of a car park or the external highway system. May also be described as a 'vehicle reservoir'.

Routine inspection: regularly scheduled visual inspection of a car park structure, cladding, edge protection and any other defined aspects of the facility.

Routine maintenance: periodic activity intended to prevent or correct the effects of minor deterioration, degradation or mechanical wear of the structure or its components.

Standard design vehicle (SDV): any vehicle that falls within a set of dimensions applicable to 95% of vehicles currently in use in the UK.

Static capacity: total number of **bays**.

Structural appraisal: evaluation of the structural adequacy of a car park structure, cladding and/or edge protection taking into account its environment, likely usage, extent of deterioration and anticipated **design service life**.

Switchroom: area containing an assembly of electrical equipment for control and protection of electrical circuits.

Turning circle: diameter of the swept path by outermost part of vehicle when on maximum steering angle.

Vehicle ramp: inclined surface for use by vehicles to access parking decks at different levels, that has a gradient steeper than 1:20.

Vertical circulation module (VCM): module comprising two parallel internal longitudinal parking ramps interconnected by a transverse vehicle ramp.

Wearing screed: structural concrete applied *in situ* to concrete or precast units to act compositely with them and to form the finished flooring, as defined in BS 8204. Formerly known as a 'structural topping'.

Wheelchair accessible vehicle (WAV): vehicle that has been specially adapted to enable a wheelchair user to travel safely and comfortably while seated in their wheelchair.

Part 1
Conceptual design

Introduction

General

We are living in a time of great change to the demands that are being placed on parking structures. For more than 120 years the purpose of a car park has been to store automobiles. With the transition to electric power, car parks are already starting to serve as vehicle refuelling stations, as well as vehicle storage facilities[†].

The Institution of Structural Engineers first published *Design recommendations for multi-storey and underground car parks*[1] in 1976. Three subsequent editions have been published since then — the most recent in 2011, reflecting considerable change in car park design and the learnings from a number of serious structural failures, including the Pipers Row deck collapse (Figure 0.2).

The Institution initially intended to create a 5th edition, but with so much change since 2011 it quickly became apparent to the contributors that the guidance needed to be substantially rewritten and reformatted, to better capture the myriad considerations which go into the design, construction, maintenance and reuse of a modern car park. Hence the new title of *Car park design*. Although completely new, it is hoped that it will quickly become the *de facto* guidance for the design of car park structures, and it's existence is certainly owed to the four editions of *Design recommendations for multi-storey and underground car parks* that preceded it.

It is presented in two parts:

Part 1 (Conceptual design) covers the wide range of factors that need to be taken into account in ensuring a car park is successful throughout its design service life. It is intended for those involved in procuring a car park; including clients, architects and project managers among others.

Part 2 (Detailed design) covers certain aspects in more depth, with detailed recommendations for those responsible for inputting to the detailed design (e.g., structural and civil engineers).

Background

Vehicular design and use is changing rapidly. Major manufacturers are moving towards sole production of all-electric vehicles — and, in the UK, will be banned from selling petrol and diesel cars and vans by 2030 (and hybrid cars and vans by 2035).

The pace of change towards autonomous and semi-autonomous vehicles is also accelerating, and car parks will need to be designed to accommodate this. Car sizes and weights are increasing and modes of transport are also changing — with electric cycles and scooters becoming mainstream, and a significant increase in traditional bicycle use.

In 2019, the Structural Engineers Declare movement was launched in an effort to raise awareness and effect changes in response to the climate emergency[2]. The IStructE pledged to support the movement as part of its commitment to assist the profession in tackling the climate emergency.

A number of serious fires have occurred in car parks which were designed to current guidance. It is clear that in certain circumstances, the guidance may not produce a fire-safe design. Additionally, we are now exposed to risks from vehicle compositions and propulsion systems which current guidance does not acknowledge.

We have moral and legal duties to make buildings accessible for all — and car parks are no exception. In many countries including the UK, people are living longer and are more independent for longer. The physical impairments of

[†] In fact, this is nothing new; the first parking structures built in the UK more than 120 years ago were specifically designed to store and service electric vehicles. The oldest known, was bult in London in 1901 by the City & Suburban Electric Carriage Company. It had seven floors and could accommodate 100 vehicles. The structure no longer exists but Figure 0.1 shows the interior of a similar facility.

Figure 0.1: Niagara Garage (1913), Westminster, London, UK

Figure 0.2: Punching shear failure: Pipers Row car park (1997), Wolverhampton UK

a greater number of car park users need to be born in mind. We must also address relevant mental health aspects, as these are often overlooked. *Car park design* attempts to address these issues — some of which can often be accommodated easily with thoughtful design. Sadly, it is also necessary to consider physical measures to deter suicide attempts. Chapter 16 is dedicated to this.

While the contributors to this guidance cannot predict the future, car parks we build today will be designed to be in use for many years, and will need to remain viable and useful in a rapidly changing situation. *Car park design* has been written with this in mind.

Scope

The scope has been widened, and covers aspects of construction, maintenance, reuse, and demolition which impact the initial design.

These recommendations are intended to be set at a level of good practice. The guidance does not set a gold standard, and a particular design may want to go beyond the recommendations in this guide. Conversely, standards and national regulations may give recommendations at a lower level. This guide is intended to represent the state-of-the-art, and to encompass changes to vehicles and use of car parking areas that other standards may not have considered. The recommendations should therefore be interpreted as those necessary for good practice.

The guide is intended to be useful to all who are involved in the design and construction of car parks, not just structural or civil engineers. Part 1 gives a broad overview of the issues to be considered at conceptual stage, and is accessible to those without a technical background.

The guide is relevant to all car park structures, including standalone multi-storey car parks, underground car parks and those within buildings with other uses; such as office, residential and retail. It is also intended that the guidance is applicable to large car parks and to all those accessible by the public.

It is not intended to cover car parks at grade, but some of the guidance will be relevant nonetheless. Small private car parks may have different operational standards and it may be appropriate to deviate from some of the recommendations made here.

Chapter 20 describes, in general terms, some of the types of car stacker systems available. However, these incorporate lifting, mechanical and robotic systems and are considered to be proprietary. Advice should be sought from the manufacturers.

Car park design complements existing standards and codes of practice by offering design guidance that is specific to car park design and construction. It is not intended to be used in isolation and intentionally refers to current codes in both the UK and mainland Europe without repeating the details they contain.

Designs may require consideration of specialisms such as seismic design. Where this is necessary, the designer should take specialist advice. The guidance does illustrate areas of special concern in car park design where these differ significantly from standard practice as set out in building and construction codes. The principles are felt to be applicable worldwide, although it must be recognised that local, regional and national variations to design requirements exist and these should be considered in developing the design basis.

Car park design is not intended to provide the definitive approach in any situation; as in all circumstances those best placed to decide on the appropriate course of action will be the engineers and architects undertaking the particular project. This guidance is applicable to most design scenarios but specific advice from a recognised subject expert should always be sought should the circumstances require it.

1 Developing the brief/the design process

1.1 Introduction

This chapter has been written to aid discussions between the client who wishes to commission a design for a car parking facility, and the structural engineer and/or architect. As the design progresses, early proposals will be revisited and refined until a final working solution is arrived at.

The flow chart at the end of this chapter (Figure 1.1) shows each step in the process; from initial concept to dealing with removal and disposal of the structure at the end of its useful life.

It is advisable to carry out both technical and financial feasibility studies before proceeding with a detailed design so that the viability of the project can be tested, and all the key requirements established. This will require a multi-disciplinary team of architects, planners and engineers.

The process of developing the brief is described from beginning to end, from identifying that a parking facility is required, through to sizing, design, operation, maintenance, and decommissioning and disposal. Subsequent chapters describe each step in detail.

At the outset, the brief will be a statement of the client's requirements, but as the site's constraints, and the planning authority's restrictions become better understood it may have to be modified several times. Similarly, the design will be adapted and refined as more information becomes available.

The demands placed on a parking structure are continually evolving. As cars develop, so too should the building that is designed to house them. Typically, a car model will have a life expectancy of about a decade before there is a major redesign. Historically the trend has been for car sizes and weights to increase. By comparison a car park can be expected to last for 50 years or more. It is important to make the brief as flexible as possible so as to be able to accommodate future vehicle designs. This is particularly relevant now, as the world transitions from the internal combustion engine to electrically-powered vehicles.

Inevitably, as the designer attempts to satisfy a series of requirements and constraints that may well be incompatible, the process may be both circular and iterative until an optimal solution is developed.

1.2 Health and safety

The importance of health and safety cannot be over-emphasised. It is central to the whole process and must be constantly referenced and validated.

It is essential to protect the health, safety and welfare of those engaged in the construction, use, maintenance and demolition of a project. This can be achieved by compliance with the Construction (Design and Management) Regulations 2015 (CDM)[3] which take effect at the start of the conceptual stage and run throughout the project. The client (or a CDM coordinator appointed to advise the client on health and safety during the initial design and planning stages) is required to gather the pre-construction information used to identify potential hazards, and to plan for the safe use of the site.

At this early stage, the client or CDM coordinator should appoint a principal designer to develop the conceptual design. Note: the design personnel may change should the project become a 'design and build' project.

1.3 Understand the need/establish the size

Almost without exception car parks are not destinations in themselves. They service the parking needs of an associated building, facility or area. It is necessary to first understand the primary need/activity, in order to determine the size of the ancillary service (the car park) you wish to produce.

The size may be influenced by several factors which are not necessarily mutually exclusive e.g.,:

- Client's initial requirement (e.g., to address a shortfall in provision)
- Demand — generated from size and anticipated use of building/s to be serviced by the car park
- Local planning constraints
- Site constraints
- Relevant data obtained from other sources (e.g., TRICS database[4])

The client's requirements may be derived from previous experience of similar developments, a corporate strategy, or for some other reason. However, the design team may also wish to draw on external data if a balance between client requirement and practical engineering solutions needs to be struck.

There are numerous traffic studies available, many of which have been incorporated into the TRICS database. Note that TRICS is not a parking demand data source. It is a collection of more than 7,000 studies used to assess the traffic impact of various developments. However, by using TRICS data it is possible to estimate how many vehicles might arrive and depart from a site, and hence estimate vehicle accumulations by assuming how long vehicles will stay on the site. It will also allow the design team to understand likely peak flows — aiding the assessment of entry and exit capacity requirements.

'Parking accumulation' studies typically link a car park to a type of building use and so offer guidance on maximum accumulation and hence the size of the facility required. It is also important to be able to estimate bay turnover, as this information will be required to properly design the internal circulation pattern.

It is important to understand that the relationship between the building size and parking requirement may not be linear. If an office block of Xm^2. requires Y spaces, it would not be unreasonable to assume that a similar building of $2Xm^2$ requires twice as much parking to meet its needs. However, for other uses, such as retail, the relationship is less clear. Staff numbers may be proportional to the size of the shop, but customer numbers may not. Such cases would benefit from a more detailed study, rather than simply scaling numbers up or down pro-rata to the floor area.

If no data exists from previous studies, the designer can generate an estimate of likely requirements from first principles based on the size of a building and its use/s.

Example

Consider an office building of 10,000m^2 gross floor area.

If, on average, each worker occupies 14m^2, the building will be occupied by:

10,000/14 = 715 staff

There may also be ancillary staff e.g., cleaners/caterers/security who do not occupy a desk space, so assume that total staff in the building will be 750.

Not all 750 will arrive in their own car of course; some will use public transport, or cycle, or walk, or come as a car passenger. Assume 15% by public transport and 5% walking or cycling, so 80% come by car. However, some people will arrive at work as car passenger; let us assume that average peak hour car occupancy will be 1.1. So, worker parking requirements would be:

(750 x 0.8)/1.1 = 545 spaces

Almost all buildings will have visitors, and so the capacity should also allow an appropriate number of spaces to accommodate these, unless there are suitable public facilities nearby that can accommodate this demand. This will need to be estimated on a case-by-case basis.

For this example, assume 50 visitor spaces, giving a total need for 595 spaces. However, it is not good practice to provide just enough spaces. A reserve should be included to deal with uncertainty. Established practice is to recommend that the car park should have 15% reserve capacity.

This gives a demand for a 684-space car park.

In the example of an office block where demand may be more predictable and steady perhaps a smaller reserve would be appropriate.

If the activity in the building includes employees working to shift patterns, especially if shifts overlap, then once or twice a day there could be two shifts of workers on-site at the same time. This impact could of course be managed by staggering shift change times for different parts of the workforce. The car park must be sized to accommodate this.

Commuter car parks, for example at train stations, are also likely to experience this type of tidal flow, compared to a retail car park where people are likely to be coming and going all day with a shorter stay duration.

Clients will likely need to prepare a travel plan setting out the proportion of staff/visitors using X,Y,Z mode of transport, both for the existing situation and for the organisation's intended objectives. This document can serve as a useful resource for confirming parking demand and is normally a mandatory part of the planning submission for any new parking facility. As well as car parking, it will also cover drop-off requirements, bus stops, car-sharing initiatives, electric vehicle (EV) and cycle parking, many of which may be accommodated within the car park development.

Generally, local plans seek to promote more sustainable forms of transport and reduce car dependence. In the UK, local plans will typically carry maximum standards setting out a not-to-be-exceeded ceiling for how much parking may be provided for a given development. These standards will either be in the form of "X spaces per Ym^2 gross floor area" or "X spaces per unit", where the proposed development can be better described in terms of units, e.g., housing. Usually, the local council will also provide guidance on the proportion of accessible parking bays and possibly, parent and child bays required at different facilities. If the design team's estimate of parking requirements varies significantly from the local plan standard, it will be necessary to reach an agreement with the local planning authority (LPA). If the LPA is not willing to meet the developer's perceived needs, the viability of the project needs to be considered. Table 1.1 illustrates a typical council's parking standards.

Inevitably, the given site and any associated site constraints may determine how much parking can be accommodated. The footprint and topography of the site will determine how many spaces can be fitted per floor, with allowances for entry and exit points, staircases, lifts, ramps and any plant rooms or ancillary services. In addition, there will often be a height limit set by the LPA, either pre-defined in planning policy or negotiated through the planning process, which may in turn limit the number of above-ground floors possible. Access to the highways network can also constrain the parking capacity that can be provided on a site – for example, an existing junction may have a cap on the number of vehicle movements considered acceptable for reasons of safety or congestion. These constraints can be identified through early engagement with the local highway authority.

Table 1.1: Parking standards used by UK councils

Development	Description	Maximum number of parking spaces
Commercial	B1(a) office	1 space/30m^2
	B1(b)/(c) high tech/light industry	1 space/45m^2
	B2 general industrial	1 space/45m^2
	B8 warehouse	1 space/90m^2
	B8 wholesale cash & carry	1 space/30m^2
Retail	Non-food retail and general retail (covered)	1 space/20m^2 covered area
	Non-food retail and general retail (uncovered)	1 space/30m^2 uncovered area
	Food retail	1 space/14m^2 covered area
	A2 financial/professional services	1 space/20m^2
	Garden Centre	1 space/25m^2
General residential	1 bedroom units	1 space per unit
	2–3 bedroom units	2 spaces per unit
	4+ bedroom units	3 spaces per unit

1.4 Consider other uses

Many multi-storey car parks are built as free-standing single-purpose above-ground structures, and so there are no other constraints on the design from other activities. There may be other uses within the structure, such as motorcycle or cycle parking but these will not impose any additional design constraints on the structure, apart from maybe the positioning of entry and exit points for these vehicles. Two-wheeled traffic should be kept separate from car traffic if possible, and in particular should not be required to use the vehicle ramps.

However, some car parks will have other activities co-located in the structure. For example, there may be retail units on a ground floor street frontage or occupying the whole of the ground floor. There could also be a restaurant or similar facility at roof level, taking advantage of the vista offered from the top of the car park (Figure 1.2).

Any ancillary use will directly impact the design. Retail spaces tend to have greater floor-to-ceiling height than a car park deck and so this would particularly impact the ramp length needed to rise above the retail. There would also be additional requirements for waterproofing the lowest deck of the car park, to ensure that none of the polluted water that will be discharged on to the deck will penetrate into the retail area.

In addition, fire safety standards will be different if other uses are included in the structure.

If upper-level retail is a factor, it will also add extra design requirements. The retail unit would require separate, free-standing customer access. Both stairs and a lift would be required. In addition, consideration has to be given to servicing the facility and providing utilities. Car parks do not have gas services, but a restaurant may use gas for both heating and cooking. If a gas service is to be provided, great care must be taken to ensure that it is protected from any risk of collision by a car. Also, the car park structure needs to be designed for the potential risk of disproportionate collapse resulting from a gas explosion (Section 11.1.9).

Car parks that exceed six storeys in height are a higher consequence class with respect to disproportionate collapse and require a systematic risk assessment in accordance with Approved Document A (Part A3) of the Building Regulations[5]. The IStructE's *Structural robustness and disproportionate collapse in buildings* guidance also provides relevant information[6].

Figure 1.2: Ground floor retail units and glass-fronted rooftop restaurant

An underground car park will cost about three times as much to build, compared with an equivalent facility above-ground. They require:

- More forced ventilation
- More lighting
- Enhanced fire protection
- Pumped drainage

adding to the whole-life operating cost of the facility.

However, site constraints and the developer's desire to utilise all the available land for other activities may make an underground car park the preferred solution. This means that the car park has essentially become a part of the foundation of the building that it has been built to service.

Large buildings are generally built using a grid of columns, and the grid size will have been chosen to optimise the use of the floor space in the building. For example, a hotel might typically be 6 x 10m or 6 x 8m. If this constraint is extended downwards into the car parking, it is not conducive to a good layout or a spatially efficient design. One solution would be to utilise a transfer slab, to allow a different grid spacing above and below the slab, optimised to the activity in that part of the building. However, this is expensive, and the car park designer is often left to make the best of a sub-optimal grid.

A car park may require enhanced fire suppression to protect the other uses (e.g., retail) in the event of a fire (Chapter 7). The design brief needs to fully identify the requirements that the fire suppression system imposes, and to ensure that they are included in the design, especially in relation to water supplies and tanks.

In addition, the overall design may place elements of the building's services in the car park. The car park designer needs to consider both the positioning of these plant rooms and the need to service them in the future; with particular care being taken to ensure that the headroom and access route is adequate for larger servicing vehicles.

1.5 Determine site geometry and constraints

If the car park is free-standing, rather than part of a larger development, its orientation will be determined by the shape of the site and what provides optimal land utilisation. Chapter 5 gives the recommended minimum dimensions for bay size and aisle widths. The floor plate design should be oriented to make best use of the available land. If the site is sloping, this may dictate the type of layout (Chapter 5).

In orienting the car park the designer also needs to consider internal circulation, positioning of ramps and the vehicular connection/s between the parking and the surrounding highways. The number of entry and exit lanes will be determined by the anticipated peak entry and exit flows. Depending on the type of access control selected (Chapter 4) it will be possible to estimate how many lanes will be required. For example, a tag system would allow upwards of 350 vehicles per hour per lane to pass, so the number of lanes required would be determined by dividing peak flow by 350. Inevitably equipment will fail from time to time and so it is good practice to provide a reserve so that if a lane fails at a peak time, the required level of service can be maintained. If the entry and exit lanes are adjacent to one another it may be possible to reduce cost by having a two-way reversible lane which can be switched between entry and exit as required.

The car park will also require provision for office/service space and for plant rooms to accommodate utilities, computer hardware etc. Even if a car park is to be operated without staff there should be a basic facility so that service staff attending the site have somewhere to operate from. As a minimum the building will require storage facilities for cleaning materials, and any equipment used in the car park. If a car park is to be staffed, a secure office facility is required, possibly including a reception area for customers.

The car park will need plant rooms for lift motors, ventilation and electricity supply boards. With the imminent conversion to an all-electric vehicle fleet, there will be a need for infrastructure to support the charging facilities. This may include a local electricity substation.

The layout will require pedestrian facilities for vertical movement and fire escape. In the UK, regulations specify a maximum walking distance to a protected fire escape. The designer must ensure that this can be achieved.

In positioning the fire escapes the designer needs to carefully consider the escape route from the bottom of the escape stairs to clear the building. The design of the fire escape must ensure that users are protected from smoke and fire. The calculation of the correct size of stairs is explored in Chapter 14.

Separate from fire escapes there may be stairs and lifts to allow drivers and their passengers to move vertically under standard conditions. Chapter 3 gives general guidance. In a smaller car park these facilities would be located close to the point nearest to where people would want to go after leaving the car park. In a larger car park, where people may wish to travel to multiple places outside the car park, multiple lift and stair cores could be considered, in order to provide a better service.

The designer will have a layout for a floor plate accommodating all the required ancillary infrastructure and parking spaces. From this, the number of floors needed can be estimated. However, this would have to be considered in the context of any height limit imposed by the LPA. In extremis it may be necessary to build down as well as up with some parking provided below ground level in order to provide sufficient spaces.

Historically in the UK, most above-ground car parks have been built without a roof. This means that during icy weather, the top deck becomes unusable, with an attendant loss of capacity and revenue. In addition, because the car park is open to the weather the structure becomes more susceptible to damage from the elements. To moderate this damage, car parks have waterproof membranes applied to the upper storeys. This increases maintenance costs as they require replacement approx. every ten years.

However, if a car park is designed with a roof:

- Water ingress and damage is largely avoided
- A waterproof membrane is no longer required
- The top deck is always useable
- The roof provides an opportunity to install photo-voltaic panels, for power generation
- By enclosing the top deck of the car park with a roof and suitable full-height cladding or enclosure, the risk of suicide by jumping will be reduced

If the car park is to be a part of a larger development, the designer should carefully consider the relationship with other aspects of the development and how it connects to the highway system and pedestrian routes to determine best position on the total development footprint. Automatically placing the car park in the more remote part of the site at the cost of lengthy access routes and extended walking distances is poor design.

Particularly in an urban environment, the site may have sub-surface utilities. These would need to be identified by a combination of research with the local authority and the utility companies, and direct site surveys. The designer needs to be aware of any pre-existing pipes, ducts or cables and plan accordingly. It may be possible to divert some utilities but the cost of moving others, such as high voltage electricity or fibre-optic telecoms cables, is likely to be prohibitive and so the car park would be better designed around them. The design may need to cater for future access to services that remain within the car park's footprint.

The development may require an archaeological investigation. This may simply require site investigation followed by the documenting and removal of artifacts. However, in extreme cases, when the remains of an exceptional building are discovered, the local authority may require that the foundations of the new building are modified so that the earlier building is bridged and preserved for future generations.

If the site has previously accommodated burials, it would normally have to be fully excavated and the human remains removed and reinterred elsewhere.

The design team will require a detailed knowledge of the existing ground conditions in order that the structure can be properly designed. This will require a detailed survey of the ground composition and the water table.

On a brownfield site the design team also need to be aware of any soil contamination and to design appropriate remedial measures to either remove or contain the contamination. Generally, all sites will require a contamination survey.

The design team will also need to carry out full arboricultural and ecological surveys, to establish what trees and ecological habitats are present and, if appropriate, what steps are necessary to protect and preserve them.

An unexploded ordinance survey may be required where the site has been assessed as being at risk.

The full list of surveys will usually be set out in the LPA's validation checklist, including some (such as a ground investigation) which may be required to inform the developer's cost plan and viability assessments.

1.6 Planning and Building Regulations

In the UK, under the Localism Act 2011[7] and the National Planning Policy Framework[8], the LPA is required to set out its 15-year vision for the future development of the authority area in a 'local plan'. This document is prepared in the context of both regional and national planning policies. Not all local authorities have completed this process and due to the need to update this document on a regular basis, there may not be an up-to-date local plan in place at the time of a development. The Plan's significance – or 'weight' in planning terms – decreases with age. Therefore, it is important to review the current and emerging planning policy and to discuss the planning context with the LPA prior to submitting a planning application.

The plan is unlikely to determine exactly where or how much parking should be provided within the LPA's area, but it may be part of a suite of policies including neighbourhood plans, area action plans, proposals maps or supplementary planning documents which are sometimes more prescriptive.

These documents are a plan for future development/redevelopment with indicative land uses and a timescale, depending on local conditions. The plan could be very prescriptive as to the exact site for a facility and the constraints that would be applied to any proposal. A modern steel-clad structure that may be completely acceptable in a rapidly developing town centre would be completely unacceptable in the centre of a Georgian streetscape. The architectural and material constraints placed on a new car park will be subject to the local situation and the expectations and opinions of local decision-makers.

Most local authorities have now adopted Community Infrastructure Levy (CIL) policies[9] (national regulations dating from 2010 designed to fund infrastructure) and as such, there may be additional charges to be taken into account by a developer. The CIL charge is based on development typology and location, and is chargeable as a rate by m^2 of floor area[†]. All of these criteria vary by local authority and must therefore be checked against local policy to confirm any charge due. Developers must also consider charges that may arise from Section 106 and Section 278 Agreements (Section 1.11) that could be due as part of their project.

Any new building will be required to conform to current Building Regulations[10] (as may alterations to existing buildings and changes of use). These set out mandatory minimum requirements for many aspects of a building pertaining to:

- Health and safety of persons in and around the building
- Conservation of fuel and power
- Accessibility
- Security

Building Regulations are regularly updated and now specify the need for (and numbers of) electric vehicle charging facilities to be provided in new buildings.

International building codes set similar requirements, and the regulations applicable to the location of the building should always be used.

1.7 Sustainability

It could be argued that a car park is axiomatically unsustainable in that it facilitates the continued use of personal transport and the mobility that this supports, rather than promoting more sustainable modes of transport and active travel such as walking and cycling. However, the reality is that not all trips can be served by more sustainable modes.

[†] CIL is likely to be superseded across England by the Infrastructure Levy (IL) — part of the Levelling Up and Regeneration Bill. This will take effect over the next decade and will be mandatory for charging authorities to implemant (unlike CIL, which is optional and varies by authority). It is calculated as a percentage of the final gross development value, with rates and thresholds set by each charging authority. Developers should take specialist advice to understand the potential liability of any new car park project.

Throughout the world, governments are moving to phase out the fossil-fuelled internal combustion engine and move towards a future where most cars are powered using rechargeable batteries. To be fit for purpose for the longer-term, car parks need to be designed to accommodate solely battery-powered vehicles. However, this will have significant implications for future design:

- On average a battery-powered vehicle is approx. 30% heavier than a traditional model fitted with an internal combustion engine. This becomes important with larger vehicles where the total weight of some future vehicles will exceed current design advice (Chapter 11)
- The deck bin width needs to be sufficient to accommodate battery charging facilities. This would add to the span required. These two factors taken together will impact on the deck design
- Ventilation in car parks is currently designed to remove both the heat from cars, and the noxious fumes, principally from engine exhaust. With the cessation of vehicle exhaust fumes, ventilation requirements could be reduced considerably. Although this would undoubtedly save operational energy in standard use, it should be noted that there may still be requirements for smoke ventilation systems (Chapter 13)

1.8 Accessibility

The designer must ensure that any design meets local planning and Building Regulations requirements, and enables their client to fulfil their duties under The Equality Act 2010[11] and (for public authorities) the Public Sector Equality Duty.

1.9 Futureproofing

This guidance has been based on the best information available at the time of writing. However, once built, a car park can be expected to be in use for a minimum of 50 years. Over five decades, automotive design will change dramatically and unpredictably; just consider the cars of the 1970s compared to today's models.

A fundamental change is emerging as a rapidly-increasing number of electrically-powered vehicles enter the market. Hydrogen fuel cells and hydrogen-powered internal combustion engines are also being developed, although they are at a much earlier stage of evolution.

The UK Government has announced a programme for phasing out fossil-fuelled cars. This will alter the function of a car park; from being solely for vehicle storage, to a place to store vehicles as well as a refuelling facility — with the attendant need to provide battery charging facilities *en masse*. The market for battery charging technology is not yet mature; with several different designs of connection and different charge rates available. To futureproof the facility, any battery charging system should be easily replaceable if a particular design becomes dominant or redundant. Currently, most electric cars are plug-in, however, work is already underway to develop induction charging technology, so that vehicles are charged from equipment buried beneath the surface of the parking bay. The car park designer needs to be aware of and cater for changes in charging technology.

In early parking structures the design often used short span beams with inboard columns. The spacing between these columns imposed a limit on bay widths and over time, as vehicles became wider, bays became too small. For example, the 2021 VW Golf is 18cm wider than the original model, and this difference is amplified by the increase in the thickness of the doors, which now contain:

- Anti-penetration bars
- Airbags
- Speakers
- Electric motors to operate windows and mirrors

The long-term trend for cars seems to be to gradually increase the dimensions of the bodyshell. This has meant that in some early car parks with internal columns, with three parking spaces between columns, the deck has had to be re-marked to provide just two spaces. If a car park is designed without internal columns, then it does mean that, as cars become wider, bay markings can be adjusted accordingly.

There is also work being undertaken to develop autonomous (self-driving) vehicles, with both motor manufacturers and IT companies actively testing prototype vehicles. Indeed, there are already autonomous vehicles in limited use.

An autonomous vehicle means that the control functions traditionally performed by the human driver are provided by a control computer within the car. It seems probable that any car park being built today will be used by autonomous vehicles (Chapter 10).

A conventional car park is a special-purpose structure built for a very particular use. Alternative uses in the future could be limited by features like sloping floors (for drainage) and the lack of natural light. The potential for repurposing a redundant car park structure, rather than demolishing/disassembling it, is discussed further in Chapter 19.

1.10 End of life

There will come a time when the structure is no longer fit for purpose. This could be because the design of the motor vehicle has evolved to the point where it does not fit with the building's dimensions, or because the building has deteriorated. It may also be that the building is no longer needed. At this point the building may be demolished and removed to allow re-use of the land. However, it could alternatively be adapted for another use.

When designing the car park and considering the construction methods and materials, the designer should also consider the end of life and the materials that are to be utilised and the potential for components to be reused or recycled[12].

Car parks are typically built using steel or reinforced concrete. Steel car parks tend to be utilitarian in their internal design but can be clad in a large range of external finishes giving a range of designs (Figure 1.3). The obvious advantage of a steel construction is that the steel components can be disassembled, and the material can be re-used, either as components for another building, or scrapped and recycled.

Concrete structures are heavier in appearance but perhaps better suited to provide a more traditional "solid" building form (Figure 1.4). The remaining concrete can be processed and used as an aggregate.

BS 8500-2[13] describes the standards for recycled aggregate consisting principally of crushed concrete. However, care must be taken when using material obtained from certain parts of a car park structure as it may be contaminated with chlorides from road salts.

It is also common for car parks to be built compositely; comprising a steel frame and precast prestressed concrete planks with a wearing screed.

Figure 1.3a: Steel car park: interior finish

Figure 1.3b: Steel car park: exterior finish

Figure 1.4a: Concrete cark park: interior finish

Figure 1.4b: Concrete cark park: exterior finish

1.11 Vehicular traffic flows

The designer must understand how the facility will connect with the surrounding highway network, and ensure that the flows of vehicles entering and leaving the car park can be accommodated without causing congestion. If accommodation isn't possible, the developer may be required to fund appropriate highway improvements (commonly known as Section 278 Agreement costs in the UK) in addition to employing a traffic engineer. These costs should be included in the project's budget.

The design team will need to know peak vehicular movements and timings and will likely use suitable traffic simulation software to model the surrounding critical junctions.

The designer must ensure that they design the correct number of entry and exit lanes to accommodate the anticipated flow.

For entry, the designer also needs to understand how vehicles using multiple entry lanes will merge inside the car park and distribute round the structure to find parking spaces. The design must ensure a vehicle reservoir of sufficient capacity for the flows to merge without blocking. If this area is not adequate, traffic will gridlock inside the car park and this in turn could block the entry lanes and lead to traffic queuing back onto the highway. This must be avoided.

Immediately before the barriers (i.e., at the edge of the highway) the designers need to consider both the interaction with pedestrians crossing the entry/exit, who have right of way, and the sightlines for vehicles exiting the car park – particularly those turning right, to ensure that the movement can be made safely. For entering traffic, particularly those turning right into the car park, the highway needs to be of sufficient width to allow a queue of vehicles to stand while waiting to turn into the car park without obstructing the free flow of traffic.

Similarly, the number of exit lanes should be commensurate with the peak exit flow, again with a reserve capacity if possible. The connection with the highways needs to be carefully considered so that vehicles exiting the car park can swiftly join the highway and not block the exit lane.

Some larger car parks may have multiple entry and exit points on the car park's perimeter to make access to and egress from the car park easier. Where this is the case the allocation of entry and exit lanes must be commensurate with the anticipated traffic flows.

Where entry and exit points are adjacent, the need for reserve capacity can be met by providing a bi-directional lane, which can be switched between entry and exit as required. As regular users of the car park may be uncertain as to the directionality of this lane on a given day it must be clearly signed, in advance, to allow safe merging.

When the lane is not in use, it is advisable to use signage similar to that used on highways to denote a closed lane (Figure 1.5). Experience has shown that if a red traffic signal is used, a driver may misinterpret this to mean that they must wait for the lane to clear and so pointlessly queue at a closed lane.

Car parks can operate with a barrier at entry and exit, or with free-flow lanes. Free-flow lanes could be used:

- If the operator does not wish to control access to the site
- If the car park is going to be operated with pay and display parking meters
- If the car park operator is using Automatic Number Plate Recognition (ANPR) or another form of vehicle identifier, like Radio Frequency Identification (RFID) tagging, to record vehicles

The distinct advantage of an open access system is that, because there is no impediment to flow, a larger number of vehicles can pass within a given time. The disadvantages include the need for an enforcement/management system and the inability to stop vehicles entering the car park and fruitlessly searching for a space when the car park is full.

Alternatively, a barrier-based system may be installed to control access. This allows entry to the site to be controlled, simplifies the collection of revenue (if appropriate) and can be used to restrict entry if the car park is full or closed.

Figure 1.5: Lane closure signage

The choice of access control system will depend on the operator and the clients they expect to attract. There are many different technologies and philosophies that can be used to control access, both with and without a physical barrier — with throughput capacities ranging from less than 200 to more than 500 vehicles per hour (Chapter 4).

Note: figures given in Chapter 4 are given for guidance only. Irrespective of the control system selected, the supplier should be asked to confirm the exact performance of their equipment under the conditions in which it is going to be installed and used.

1.12 Pedestrian movements

The number and position of pedestrian access points will be determined by how the car park connects to the surrounding buildings and activities. This will also determine the number and position of pay points. Historically most car parks have had little or no security at the pedestrian entry points. However, some modern ticketing systems can allow the parking ticket to also be used to operate the pedestrian entry door, and so increase security by limiting access to the car park to those who are parked there.

In the UK entrances must be step-free, and any door opening requirement must be fully accessible, to satisfy the requirements of the Building Regulations.

When designing the car park and identifying the location of pedestrian entry and exit points, the designer should ensure that there are adequate pedestrian facilities to allow users of the car park to move between the car park and their ultimate destination on foot and by wheelchair. This could require the provision of pedestrian crossings, refuges, dropped kerbs or footways on the highways adjacent to the car park.

1.13 Building Services

The design must demonstrate consideration of all the ancillary equipment and services for the building's operation. Each of these requirements are addressed in detail in later chapters. They include:

- Electrical power; for lighting, ventilation, lifts and particularly for charging electric vehicles (Chapters 6 and 13)
- Water supply: for cleaning. There may also be a sprinkler system. In addition, if there are staff facilities or toilets, a drinking water supply will be required (Chapter 6)
- Drainage. If the car park has a roof, stormwater can be drained separately. However, there will also be a need to drain and handle polluted water run-off from the parking decks via a system equipped with petrol and oil interceptors (Chapter 11)
- Ventilation. This requirement will depend on the design of the building. If the car park is above-ground with an open aspect, sufficient air movement may be obtained from natural ventilation. However, if the structure is below-ground, continuous mechanical ventilation will be required and must be planned for (Chapters 6 and 13)
- Fire suppression. This may be required under regulations for life safety, or by the developer for property protection, and should be assessed on a case-by-case basis (Chapters 7 and 13)
- Lighting. Different levels of lighting are advised for different areas of the car park (Chapter 13). Since 2010 (approx.), car parks have moved towards LED-based lighting, both because of the relative longevity of lighting units and the considerable energy saving over other lighting systems. The designer should always seek to minimise energy consumption

  In the UK, BS EN 12464[14] is the standard relevant to lighting for enclosed car parks, while BS 5489[15] is the standard relevant to open-air structures.

  Guidance on acceptable lighting levels is also given by the European Standard Parking Award Off-Street (ESPA Off-Street)[16] which has been adopted by the European Parking Association (EPA). ESPA defines three levels of lighting:
  - Ideal level
  - Lower range which is considered acceptable
  - Level below which lighting would be deemed inadequate

  In the UK, the Park Mark® accreditation system[17] also sets out certain requirements for lighting.

Note: In the UK, for situations where British Standards are more rigorous than other guidance (e.g., for specific lighting requirements for accessible bays) the more onerous standard should be adopted.

1.14 Signage

Signage is needed for vehicular traffic circulation and for pedestrians — both for direction/wayfinding and for statutory purposes (e.g., fire escapes). Motorists should be familiar with the national system of traffic signs and road markings and it is strongly recommended that the internal traffic circulation signage should use those designs. In the UK, this is set out in The Traffic Signs Regulations and General Directions 2016[18].

In particular, where traffic flows merge or bisect, the priority (i.e., the party who has right of way) should be clearly shown using "Give Way" or "Stop" markings. Similarly, where movement is one-way, white arrows on a blue background and "No Entry" signs should be used to show permitted movements. The purpose of such signs is safety, and the use of non-standard signage should be avoided.

Pedestrian signage and wayfinding can be viewed more flexibly using designs developed for the car park. If there are multiple pedestrian exits the signage should clearly identify what destinations are reached by each exit. Separately, fire escapes should be signed using standard signs that conform to BS EN ISO 7010[19].

1.15 Wayfinding

> *"Wayfinding is the ability to understand where you are and where you want to be in a building and then to find the path to travel to get there"*[20]

Pedestrian signage should identify the path, and a good design will use visual prompts and symbols to help people understand where they are.

For example, in a multi-level car park the stair and lift landings could be painted in different colours for different floors. A driver, who may not remember which floor they parked on, will be prompted by a unique bright colour. In a car park in Chicago's jazz quarter, the operator has music playing on each floor — with each floor having a different classic jazz performer. Similarly, sign clutter can be reduced if external destinations are grouped and represented by symbols.

The client and/or the car park's location may also determine the wayfinding strategy. For example, at a hospital or university there may be a campus-wide strategy, with consistent colours and symbology used around the estate.

1.16 Floor plate design

The floor plate design should be set out to use the available space efficiently; allowing a driver to search for a parking space across the whole floor in a logical way, without the need to repeat a pass along an aisle. Cul-de-sacs should be avoided if possible — and if unavoidable, made as short as possible to minimise the need for a driver to reverse.

Chapter 5 sets out the minimum geometric dimensions that should be utilised for a well-designed parking area. These include:

- Bay size
- Circulation and aisles (one-way/two-way)
- Pedestrians, pedestrian circulation, stairwells, lifts and escalators
- Ramp positions, gradients and dimensions

It should be emphasised that the dimensions used in Chapter 5 are not strict rules. The figures are a guide and if they are followed, the car park would be capable of accommodating 95% of the currently available vehicle types, and offer the users of those vehicles an easy-to-use parking facility. Making reductions in any of these dimensions would reduce the utility of the facility, eventually reaching a point where the car park is so difficult to use that it is not fit for purpose. Similarly, increasing any dimension may increase the utility of the car park, offering a better user experience, but at a financial cost.

1.17 Building design

The design will determine the number of levels needed to accommodate the required amount of parking on the site. If building height is a planning constraint, this may require either some levels to be constructed below-ground, or the developer to compromise, accepting fewer spaces.

The desirability of providing a roof is clear. It would typically reduce whole life maintenance costs by more than the additional construction cost. The roof would increase the availability of parking on the top level during inclement weather. It would also allow the installation of photovoltaic panels, allowing power generation and giving a whole life income stream.

The exterior finish to a basic interior structure can be varied almost endlessly to match, complement or contrast with the surrounding buildings. This would be something for the design team's architect to consider.

This consideration needs to take account of factors such as:

- Need for air flow
- Fire strategy for the building and its relationship with the boundary conditions of the site
- Positioning and number of lift and/or stair towers — to provide vertical pedestrian circulation fire escape and firefighting access — and how these integrate into the external design
- Positioning of entry and exit points, both for pedestrians and vehicles
- Fixtures and fittings to be utilised and their design (in the context of the building design)
- Floor plates and their relationship with the exterior

The developer may wish to pre-determine the construction method in advance of selecting a contractor to build the car park. Construction systems fall broadly into steel and concrete frame systems, and the most appropriate may depend on many factors including:

- Relative cost of materials at the time of the development. The costs of steel and concrete can fluctuate and what offers best value for money varies over time
- Desired design service life
- Environmental and sustainability considerations
- Desired maintenance regime
- Fire strategy and required fire resistance

Alternatively, the developer may wish to keep the design open to a wider range of construction systems, such that they can allow the market to give them the best value for money. It is possible to design a new-build MSCP for a range of structural grids and systems so that the client's requirements can be protected, and detailed planning consent can be achieved prior to tendering the building and selecting a construction system.

Once the design has been set and materials and construction techniques chosen the design team can evaluate the project, estimate initial costing and confirm whether or not the proposed project is technically and financially feasible (Chapter 2).

1.18 Expected design service life and asset management

A well-designed car park will have a design service life of 50 years. In some situations, the developer or end user may require an increased design life of 60 years or greater. However, with a well prepared and implemented asset management regime and effective life-care plan (Chapter 18), the building can be maintained in near new condition indefinitely.

To achieve this, it is important that the detailed design has taken full account of the need to maintain and repair the building. For example, making sure that the drainage system is fully accessible for cleaning/rodding to remove any obstructions, and that consideration has been given to safe cleaning or replacement of cladding panels if installed.

The owner/operator of any car park has a duty under law to provide and maintain any parking facility in such a manner that it does not endanger persons while in use. This includes employees, maintenance workers, visitors (lawful or otherwise), children, etc. There are no exceptions, and failure to comply with the legal obligations may give rise to prosecution.

One tool available to the owner/operator to ensure their duty is being correctly discharged, is to commission, implement and maintain a life-care plan via an appropriate experienced and qualified structural engineer or surveyor.

Life-care is a strategic and managed approach to the inspection, maintenance and management of parking structures. A life-care plan should contain all that is known about a car park, including:

- Initial design parameters
- Records of previous owners and any change of use
- Warranties and defects
- Liability periods
- Service history; including any emergency repairs resulting from accidents, fire damage, spillages etc.
- Inspection, testing, survey and appraisal reports
- Programme for re-inspection
- Proposed repair and maintenance regimes, including costing and prioritising any repairs and upgrades
- Prediction of future serviceability and planning for phased renewal or replacement

The life-care plan should be prepared using appropriate web-based reporting software; enabling the designer to upload and share new information, such as reports of inspections, test results, sketches, photographs and records of repairs.

1.19 Planning process

The planning process is the same whether the car park is part of a larger development, or a standalone project. The planning process has five steps. The first three are optional but have the benefit of de-risking the planning process:

1. Pre application (early engagement)
2. Outline planning permission
3. Reserved matters
4. Full planning permission
5. Planning conditions

For complex proposals or where the application is sensitive, **early engagement** with the planning officer is recommended. These discussions allow for their input before the application is made public through the formal application process.

Outline planning permission, although not mandatory, allows a decision to be made on the general principles of how the site can be developed. Outline planning can identify any objections before submitting for full planning approval.

Outline planning may be granted subject to conditions requiring the subsequent approval of any **reserved matters**; these include details of the proposed development that the applicant has chosen not to submit at the outline planning stage.

Full planning permission is the decision by the planning authority to grant full planning permission. In the UK, decisions are made in line with the National Planning Policy Framework[8] and any local planning policies adopted by the local authority.

Planning permissions include **planning conditions**, some of which can be ‘pre-commencement conditions’ which must be discharged before any construction or development work begins. To discharge a planning condition requires a further submission of information pertaining to the details to be approved.

1.20 Detailed design and construction programme

Design and construction programmes facilitate the coordination of the activities required to successfully deliver the project, the sequence of the activities and their interdependence. For a programme to be effective, it must be used to plan activities, monitor progress and identify where additional resources are required.

The design programme schedules the activities and deliverables; from the engagement of the various members of the consultant team, through to the appointment of the contractor, and completion of the design.

A construction programme is a valuable management tool, enabling forward planning to deliver projects on time and within budget. Construction programmes should outline the project scope of work, sequence of activities, identify timescales, lead times, development phases and durations. The sequence of critical tasks should be identified in the programme, as this establishes the overall duration of the project.

The RIBA Plan of Work[21] is a design and process management tool that organises the process of briefing, designing, constructing and operating into eight stages with outcomes defined for each stage:

- Stage 0 Strategic definition
- Stage 1 Preparation and brief
- Stage 2 Concept design
- Stage 3 Spatial coordination
- Stage 4 Technical design
- Stage 5 Manufacturing and construction
- Stage 6 Handover
- Stage 7 Use

The RIBA Plan of Work 2020 overview provides greater detail on the application and interpretation of this process map. The IStructE's Structural Plan of Work 2020[22] complements this.

Figure 1.1: Design process and some key considerations

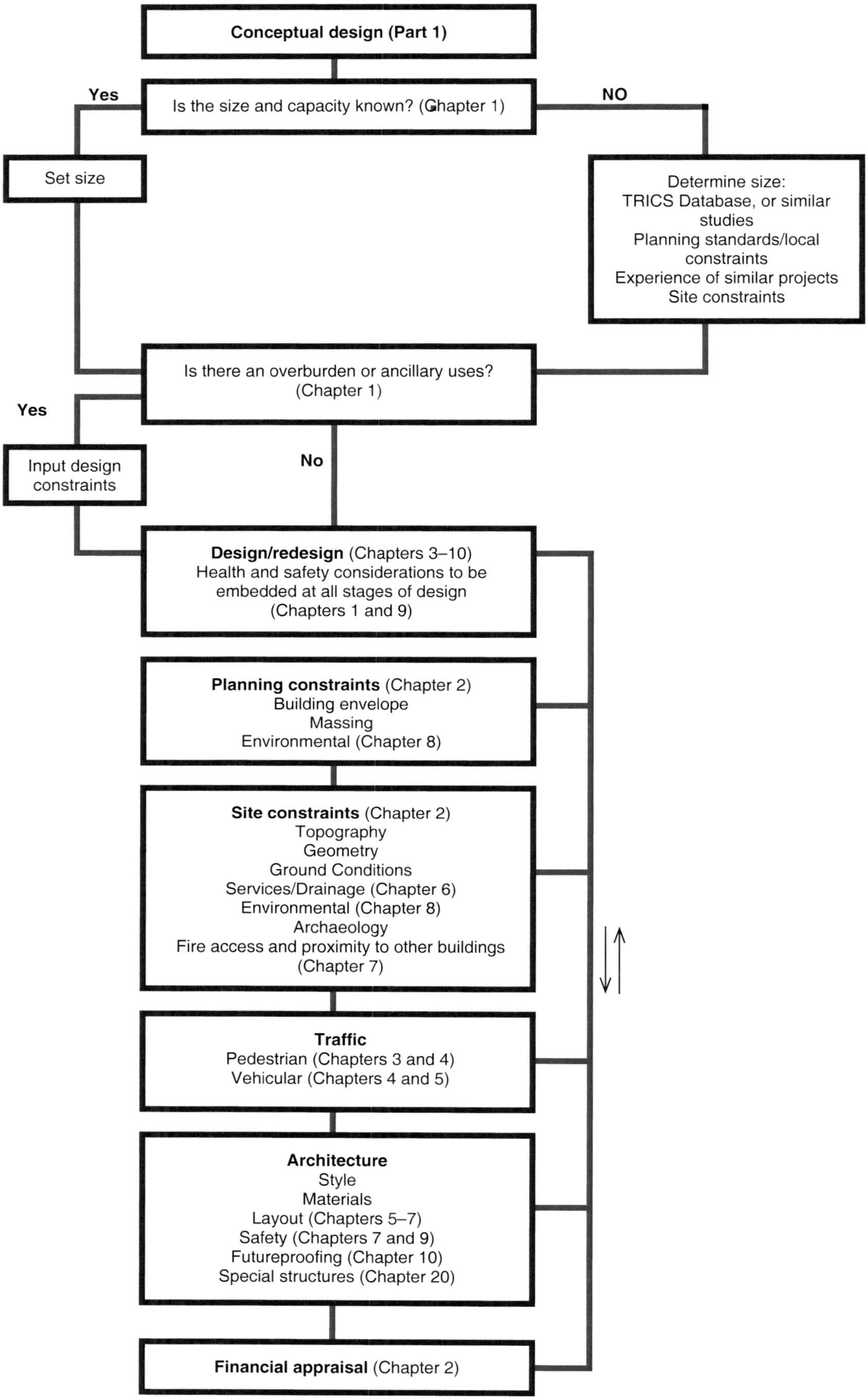

Figure 1.1: *continued*

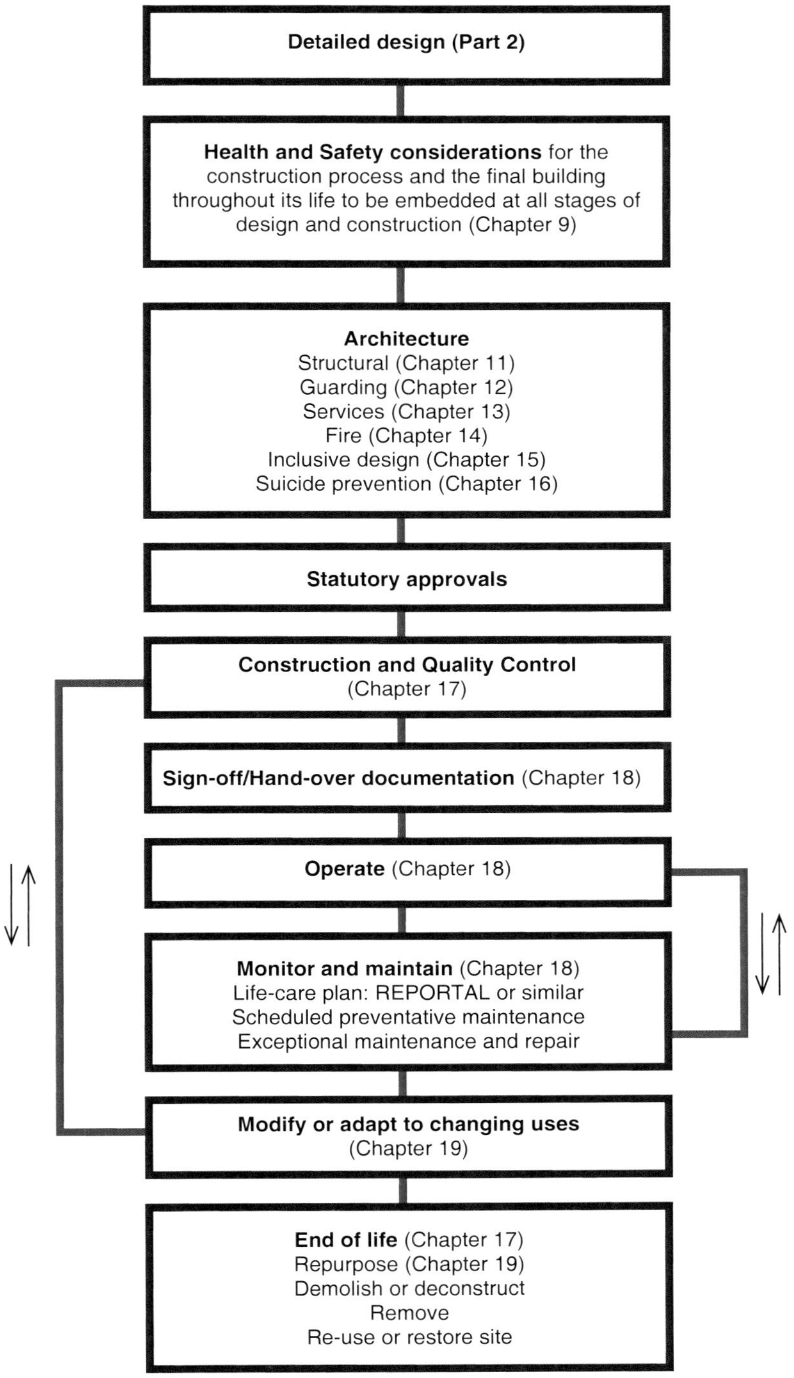

2 Feasibility

2.1 Introduction

This chapter considers the feasibility of a project. Feasibility has two interrelated dimensions: technical and financial.

The technical feasibility of a construction project depends on whether it is possible to build a structure that will provide a facility that has the functionality required. It could be argued that it will always be possible to create a structure that will meet a given need if cost is no object.

The financial feasibility of a project depends on whether the cost can be justified, based on the anticipated financial return, and/or other benefits.

2.2 Technical feasibility

It may be argued that there is always an engineering solution, at some cost, that would result in a useable facility. However, the more complex a structure becomes, the less likely it is to be viable in practical terms. To assess the viability of a project, the designer needs to consider the following:

- Can the required number of parking bays be fitted into the space available (geometry) within both the dimensional constraints of the site, and other planning restrictions (e.g., height) that apply? For an above-ground structure, if the available envelope (length, width, height) is sufficient to allow the provision of the required number of bays, the project is spatially viable. If the envelope is too small, the shortfall may be met by adding basement levels, with the consequential increase in cost and complexity.
- Do subsoil conditions on the site allow such a building to be constructed? The design team will require a properly designed and conducted investigation, to be fully informed of ground conditions. Ground conditions may be sufficient for the car park to be built with quite simple foundations, or the structure may require extensive groundworks and deep piles to provide a stable footing for the construction. Where ground conditions prohibit other solutions, it may even be necessary to consider a floating slab. Consideration of the ground conditions should also include the water table and what measures would be required to protect the car park from water ingress. This is particularly relevant if there are basement levels which might have to be tanked (i.e., where a membrane/coating is applied to interior/exterior surfaces). It may also have implications for the building's drainage requirements. It is essential that the site survey identifies any ground contamination and if found, that appropriate remedial measures are taken
- Does the surrounding highway network have sufficient reserve capacity to accommodate the traffic movements from the planned car park? The geographical extent of any traffic studies would be a matter for agreement with the local highway authority, who will wish to be reassured that the car park would not result in unacceptable traffic congestion. It is important to ensure that the proposed positioning of the entry and exit points will not exacerbate pre-existing highways congestion. If localised external congestion blocks the entrances or exits it will reduce the efficacy of the facility
- Users make as many movements into and out of car parks as pedestrians as they do as drivers. Can the car park simply and safely connect pedestrians with their ultimate destination, not just the immediately surrounding highways? The design team should ensure this is the case, and if necessary work with the local highway authority to agree highway improvements
- Can the local utilities networks accommodate the anticipated demands from the development? The water consumption of a car park is likely to be quite modest, unless the development includes other activities such as a restaurant or retail units
- The drainage system will be expected to be a 'sustainable drainage system' (SuDS), which is designed to minimise the environmental impact of water waste from the development. Initiatives such as 'green walls' and 'green roofs' would allow rainwater runoff to be used within the site (Section 8.6.3). It will also be necessary to consider flood risk and the disposal of surface water and whether an attenuation system is required

- The Equality Act 2010[11] protects people from discrimination, and places specific duties on the owner and operator of the car park. In practice, this usually means the car park must be accessible to all and must not present barriers to someone on the basis of:
    - Age
    - Disability
    - Gender reassignment
    - Marriage and civil partnership
    - Pregnancy and maternity
    - Race
    - Religion or belief
    - Sex
    - Sexual orientation

Although these protected characteristics apply equally, in a car park the most likely discrimination is towards those with impaired mobility.

- In terms of the column grid, can the design overcome any structural constraints imposed by co-located activities within the building (e.g., retail)?
- Can the design conform to necessary sustainability obligations/objectives? The project will have an impact on pre-existing flora and fauna, and the construction activity itself may impact on the water table
- Once constructed, the development will potentially produce low levels of sound and atmospheric pollution (less so with electric vehicles). Is the design able to minimise these impacts?
- Can the design utilise materials that offer durability and low maintenance, preferably throughout the design life of the building? As far as is possible the design should seek to use materials that can be recycled and reused at the end of the building's useful life[12]

Assuming that these questions can be answered positively, a quality car park can be built, and the team can prepare both a construction cost estimate and estimates for both the operating and maintenance costs which can be input to a financial feasibility assessment.

2.3 Financial feasibility

The chosen design will generate cost estimates. These should include:

- Capital or construction cost
- Operating costs (e.g., energy costs, rates, insurance, and staff costs). It would also include Parking Access and Revenue Control Systems (PARCS) costs. Note that the installation of solar panels could also generate a source of income
- Routine maintenance and life-care costs (e.g., cleaning, annual inspections and preventative maintenance). It would also include any service or maintenance contracts for ancillary equipment such as PARCS, lifts, ventilation, lighting, etc.
- Periodic costs such as lift replacement, PARCS replacement, relighting etc.
- End of life costs covering demolition and/or deconstruction and materials disposal or recycling. Note that these could also represent a significant source of income at the end of the building's life

This data can be manipulated in a suitable financial model to provide an annual cost for providing and operating the car park.

If the project is primarily a commercial undertaking, the data can be utilised to provide annual, weekly, or even daily costs. Assumptions can then be made about required occupancy levels, and the revenue required to justify investment. Comparison with rates being charged by other parking facilities in the area will determine whether the project is financially viable.

2.4 Other factors

It is important that the wider socio-economic effects that car parks bring about are understood, and that generating profit may not be the only criterion on which the feasibility of the project is judged. For example, a town centre with an under-provision of parking may well justify the investment by increased footfall and retail activity. Similarly, at a hospital, increased parking may be required to improve access to services.

The car park may also be built to be used by a building's tenants, without a direct charge for users e.g., a car park built as an ancillary use in an office development. The car park would be a necessary part of the enabling infrastructure and may contribute to rental income. The developer would need to be aware of the true cost of providing and maintaining the facility so that an appropriate charge could be included in the lease calculation for the tenants of the building.

3 Inclusive design (conceptual design)

3.1 What is inclusive design?

As long ago as 2005, inclusive design was defined by the British Standards Institution as:

> *"The design of mainstream products and/or services that are accessible to, and usable by, as many people as reasonably possible ... without the need for special adaptation or specialised design."*[23]

In the context of car park structures, this requires the designer to understand and design for both the diversity of people who may access the space as pedestrians, and the different types of vehicles that may be used by disabled people.

In the UK, with regard to accessibility for disabled people, the Equality Act 2010[11] protects not only disabled people, but also people who have an association with a disabled person (e.g., carers and parents) and people who are perceived to be disabled. The definition of disability is broad, and whether a person is considered disabled for the purposes of the Act is always determined on a case-by-case basis.

3.2 Inclusive design strategy

An 'access and inclusive design' strategy sets out how a space can be used by the likely end users. It can be used by building control bodies to determine whether the space meets the requirements of the Building Regulations, and by planners to assess a scheme against local planning policies.

An access and inclusive design strategy that identifies the anticipated users of the space, how it has been designed to meet their needs, and compliance with relevant access standards should be prepared as early as possible in the design process. It can then be updated at subsequent design stages and retained after construction, as a record of any management strategies required, and to demonstrate that a reasoned process has been followed.

3.3 Statutory background

In the design of a car park, it is important to understand:

- Planning policy: the local authority is likely to have guidance on the provision and design of accessible parking and cycle storage. This can vary significantly between different local authorities
- Building Regulations: construction of a new car park will be subject to Part M of the Building Regulations. Approved Document M[24] provides guidance on how the functional requirements can be achieved. Complying with additional guidance and standards, such as BS 8300[25] and sector specific guidance (e.g., for schools or medical buildings) also helps demonstrate compliance
- The Equality Act 2010: this does not put specific duties on designers, although it does put duties on the owner and operator of the structure once completed, as either an employer of, or landlord or service provider to, the end user. Anticipating these requirements in the design can minimise unnecessary management costs and disruption

3.4 Anticipating the access needs of users

Understanding the likely access needs of users is essential in meeting statutory requirements and minimising ongoing costs that may be placed on the ultimate owner and operator in meeting their Equality Act duties.

For car parks with a known population (e.g., staff in an office building) the current or proposed tenant is likely to have existing data that can be used to rationalise the design, though it is still important to consider visitors and adaptability for future provision.

For public car parks or car parks serving public buildings (e.g., a shopping centre), these other sources of data on local users may be useful:

- Office for National Statistics and census data
- Local authority data on the number of Blue Badge[26] holders and existing accessible parking provision
- Online maps detailing existing accessible parking provision and local facilities that may lead to a larger proportion of users with accessibility needs (e.g., hospitals)

In addition, some assumptions may be made about the types of access that need to be considered in the design. For example, it can be assumed that users driving their own vehicle have a good level of vision, and that people with reduced vision are likely to be accompanied when moving around the structure as pedestrians.

3.5 Accessible parking and the Blue Badge scheme

The Blue Badge scheme is managed by individual local authorities and intended to improve access to parking for disabled people. The Blue Badge holder is not necessarily the driver of the vehicle, and may have a variety of different access needs, including (for example) autism or partial sight. In 2022 approx. 4% of the population in England held a Blue Badge[27]. Wheelchair users represent a small proportion of this number, but less than half of the population who are automatically eligible for a Blue Badge actually hold one.

Blue Badge holders have access to different types of public parking, which varies by local authority. For example, this may include the right to park in residents' bays or restricted areas for a fixed time, or be given access to on-street accessible parking. The Blue Badge scheme does not have any statutory bearing on a motorist's right to access accessible parking within private car parks, but is often used by operators as a convenient way of managing accessible bays or improving access to disabled users (e.g., by waiving parking charges or offering extended time — as people with a disability may need more time to access their vehicle).

A motorist that does not hold a Blue Badge may still have the right to use an accessible parking bay in a private car park as a reasonable adjustment under the Equality Act 2010, if the operator could be reasonably expected to know about their disability. This is most commonly a consideration in private car parks serving a known population, e.g., a workplace, where the employer could be required to allocate an accessible bay as a reasonable adjustment, completely independent of the Blue Badge scheme.

For this reason:

- 'Blue Badge bay' refers specifically to bays reserved for use by Blue Badge holders by a local authority. These are unlikely to be present in private car parks and may not be accessible bays
- 'Accessible bay' refers to parking bays designed for use by disabled people (e.g., including access zones to the side of the vehicle) and can be managed independently of the Blue Badge scheme

3.6 Inclusive design for vehicles

Most disabled motorists and passengers, including those requiring accessible parking, use standard vehicles. However, it is essential that accessible parking and the routes to accessible parking are designed to meet the needs of a range of accessible vehicle types, including wheelchair accessible vehicles (WAVs).

WAVs are vehicles where the driver or passenger sits in their own wheelchair while travelling. Not all wheelchair users drive WAVs i.e., they may transfer into the driver's seat and load their wheelchair into the vehicle separately.

WAVs can take several forms:

- Drive-from-chair WAVs; where a user operates the vehicle from their own wheelchair (often using hand controls)
- Upfront passenger WAVs; where the user sits in their own wheelchair adjacent to the driver
- Rear passenger WAVs; where the user sits in their own wheelchair at the rear of the vehicle

WAV users may also enter the vehicle in a number of ways:

- From the rear of the vehicle using a ramp or tail lift
- From the side of the vehicle using a ramp
- Onto a platform that pivots away from the vehicle, e.g., attached to the front passenger side door

This variety of designs should be considered in the design and location of accessible parking bays, in that:

- Access space is required to both sides and to the rear of the vehicle
- Within bays arranged perpendicular to the carriageway, WAVs will usually be parked such that the rear of the vehicle is facing the carriageway
- Within bays arranged perpendicular to the carriageway, rear-loading WAVs are likely to temporarily extend into the carriageway when loading/unloading, and should be placed away from busy routes where possible

All drivers, including those using alternative controls or with restricted mobility, must meet the same requirements to hold a driving licence in the UK. Therefore, it can be assumed that WAVs move around a car park much as any other similarly sized vehicle, and that drivers of WAVs will have the same level of visibility and reaction times as any other driver.

3.7 Size of wheelchair accessible vehicles

Although additional space may be required for a user to access their vehicle, most wheelchair accessible private vehicles are converted small multi-purpose vehicles (MPVs) or commercial vehicles, and do not need special consideration. The most common WAVs currently have a height of 1,850–1,990mm.

A small proportion of WAVs may use roof-top hoist cabinets and have significantly larger headroom. Similarly, larger wheelchair accessible minibuses, such as those used by local authorities, schools or care homes, benefit from additional headroom. It is therefore recommended that a proportion of higher-headroom accessible bays, with a headroom of 2,600mm, is considered for the relatively small number of car parks serving buildings likely to be accessed by this type of vehicle. In the UK this is a requirement of the Department for Transport at certain locations e.g., train stations.

As it is often difficult to cater for this increased headroom within a multi-storey car park, provision outside of the parking structure, e.g., close to the entrances of the facility requiring this type of parking, may be a suitable alternative.

3.8 Provision of accessible parking bays

Local authorities will commonly have their own guidance or requirements on a minimum provision of accessible parking, which may be agreed as part of the planning process.

While BS 8300-1[28] recommends a minimum percentage of accessible parking bays that should be provided to serve different facilities, the provisions set out in Table 3.1 of this guidance are considered suitable, unless deemed otherwise by the local authority.

It is good practice to identify an area within the car park where groups of three standard parking bays could be converted into two accessible or enlarged bays in the future e.g., by changing the road markings.

Table 3.1: Provision of accessible parking bays

Car park used by/for	Car park size	
	Up to 200 bays	**More than 200 bays**
Employees and visitors to business premises	Individual bays for each disabled employee plus 2 bays or 5% of total car park — whichever is greater	6 bays plus 2% of total capacity
Shopping, recreation and leisure	3 bays or 6% of total capacity — whichever is greater	4 bays plus 4% of total capacity

Providing separate parking bays for larger vehicles (e.g., minibuses, campervans) and family spaces can reduce misuse of accessible bays.

Real accessible parking demand can vary significantly based on the availability of other local accessible parking and the type of facility the car park is serving. For example, a car park serving a hospital or care home is likely to require significantly greater provision, whereas a car park as part of a university campus where similar quality accessible parking is provided at grade and closer to building entrances may require very few (if any) accessible bays. Where this is the case, the rationale should be captured in the access strategy.

3.9 Drop-off points

Provision of a drop-off point close to (within 50m of) the entrance of the car park or the facility it is serving can improve overall access and reduce demand for designated accessible spaces, as it is more convenient for some users to be dropped close to the destination while their companion parks the vehicle.

Accessible drop-off points should have dimensions not less than 9m x 3.6m, to allow passengers to depart either onto a kerb or to vehicle level. Dropped kerbs and weather protection at a height of 2,600mm should be provided.

3.10 Inclusive design for pedestrians

3.10.1 Horizontal circulation

Pedestrian access routes should be designed to be accessible throughout the structure (not only serving designated accessible bays).

Where horizontal routes to accessible bays exceed 50m from the facility the car park is serving, accessible seating or rest points should be provided.

Pedestrian routes should be clearly demarcated from vehicle/cycle routes. If kerbs are provided, dropped kerbs should be arranged to avoid being blocked by parked vehicles, and be clearly visible from pedestrian routes.

In general, gradients on accessible routes within the car park structure should be no steeper than 1:20, and provide level landings at least every 500mm rise. Where gradients are steeper than 1:20, handrails will be required on both sides of the ramp to enable access. Gradients at accessible parking bays or rest areas should not be steeper than 1:60.

3.10.2 Vertical circulation

Lifts should be used as the primary means of vertical circulation. Accessible car (and cycle) parking should be located as close as possible to the accessible lift.

Where possible, more than one lift should be provided as a failsafe.

Where stairs are provided for circulation, they should meet the requirements of BS 8300, for width, rise and going, landings, handrails and visual contrast. Surface widths should be no less than 1,200mm.

Vehicle ramps should not be used for pedestrian access or egress, and are not an appropriate step-free alternative to stairs. This is the case regardless of whether vehicles are using the structure, as vehicle ramps are typically much steeper than a gradient that would be safe to use for step-free pedestrian access, and do not provide the necessary support (e.g., handrails).

Where cycle facilities are provided within the same structure, cycle routes should be segregated from both vehicles and pedestrians (typically by a separate entrance).

Pedestrian signage to indicate the location of a safe step-free route (e.g., a lift) at vehicle ramps can discourage their use by pedestrians.

4 Internal planning (and management of traffic and pedestrians)

4.1 Introduction

Planning of the car park's internal arrangement is essential to its successful operation. It affects vehicle flows, pedestrian movement and safety. Internal planning is inextricably linked to the security aspects of a scheme (Chapter 9).

4.1.1 Vehicle speed

To minimise the risk of serious injury to pedestrians, 5mph (8km/h) speed limit signs are displayed within the car park, but for the geometric design and the design of the vehicle restraint barriers a 10mph (16km/h) speed limit is used. If vehicle speed is likely to become excessive (e.g., where there are long straights) control measures need to be introduced in the form or speed retarders or illuminated signs. Raised humps or similar retrofit installations should be used with caution as they increase dynamic loading on the structure and can also be a trip hazard. The dynamic loading can be minimised by making the shape of the retarder sinusoidal (Figure 4.1) or alternatively by using rumble strips (Figure 4.2).

4.1.2 Static and dynamic capacity

The 'static capacity' is the total number of parking bays in a car park, as distinct from the 'dynamic capacity' which is the maximum inflow or outflow of vehicles (usually expressed in vehicles per hour). While the dynamic capacity is dependent on the layout (Section 5.4.7), the largest determining influence is usually the type of control employed at entry and exit, including the way in which any charges are collected.

Figure 4.1: Traffic calming: sinusoidal hump

Figure 4.2: Traffic calming: rumble strip

Short-stay parking (less than three hours) is generally associated with shopping centres, medical centres etc. which have a high turnover of bays for a given level of occupancy.

Medium-stay parking (3–12 hours) is generally associated with facilities at offices, train stations, shopping malls, hospitals etc. and has a higher peak hour flow but a lower turnover of bays than short-stay parking. Traditionally, office car parks were seen as medium-stay and would fill to 70% of the static capacity during the peak hour and empty in a similar manner, but for medium-stay parking at sports and leisure facilities, cinemas, exhibition centres, etc. the peak flow can be similar to the static capacity.

It should be noted that where several different facilities are located in close proximity i.e., shopping mall, cinema and restaurants etc. and share the same car park, the peak hour flow can be significant — especially if discharge times coincide.

Long-stay parking (more than 12 hours) e.g., at airports, has a lower peak flow and a lower turnover of bays for a given level of occupancy.

Short-, medium- and long-stay requirements are very different so the internal layout, circulation, and entry and exit requirements need to be configured taking into account the peak hourly flows and predicted turnover of bays. While the peak hourly flow rate is important, it should always be borne in mind that the area surrounding the car park may change over time, and therefore the initial design should have some inbuilt flexibility to take this into account. As a general rule, the dynamic capacity should be sufficient to permit up to 25% of the static capacity to enter or leave the car park within 15 minutes (i.e., up to 100% turnover in an hour). There should be sufficient provision for queuing on entry at peak periods (Section 4.1.6) and sufficient capacity at exit, including queueing capacity if needed to ensure that any queue does not back-up into the car park (Section 4.1.7).

In addition, when cars are arriving and departing at the same time as those already in the car park, drivers may miss newly vacated bays while searching for an empty bay. The ability to recirculate is as important as the ability to circulate the car park. Where entry is controlled, deliberate under-capacity margins of approx. 5% can be introduced to enable drivers to search for the newly vacated bays.

To maximise the dynamic capacity, vehicles entering the car park should pass as many parking bays as possible, but when exiting should pass the least number of bays. The search pattern should minimise conflict between opposing flows and have provision for recirculation should a vacant bay be missed on the entry route. Clear span structures with minimal internal columns, good visibility, good signage and bay markings enhances dynamic capacity (Figure 4.3). Poor parking discipline and bays that are difficult to enter owing to badly positioned structure or are off the main circulation route (e.g., cul-de-sacs) tend to be the least efficient dynamically and can lead to the maximum occupancy being less than the static capacity.

With minimal formalities on entry or exit, the dynamic capacity is usually determined by the capacity of the circulatory aisles, but for larger car parks the capacity of the ramping system may be the governing factor. (Section 4.1.4).

Figure 4.3: Minimal columns and good visibility

4.1.3 Aisle capacities

Determining the dynamic capacity of an aisle is complex and dependent on a number of variables which can affect vehicle speed, namely: the effective aisle width allowing for parked vehicles that overhang the parking bays and any dedicated walkways; the number of vehicles manoeuvring to and from the bays; the width and orientation of the bays; and the configuration/dynamic efficiency of the ramping system.

Typical aisle capacities for different bay dimensions are shown in Table 4.1. These capacities assume that approx. 400 cars/hour are accessing the bays and if this figure reduces, allowing the aisles to become free-flowing, the aisle capacities can be increased significantly.

Where the flows are high and there is a high turnover of bays, the dynamic efficiency of the car park will be reduced. To overcome this, the introduction of additional ramps or shortcuts to bypass the main circulation routes should be considered (Section 4.1.4).

Table 4.1: Capacities for 6m aisles with 90° parking

Bay width (m)	Bay length (m)	Aisle capacity (cars/hr)	
		Inflow	Outflow
2.7	5.0	1,055	865
2.6	5.0	955	790
2.5	5.0	905	765

4.1.4 Ramp and accessway capacities

The ramp capacity is dependent on: the width of the ramp; the number of turns and ease of accessing the ramp; interruption to exiting the ramp by the flow on the drive aisle; visibility when entering or exiting the ramp; and the gradient of the ramp. For free-flowing conditions, the operational capacity of straight ramps is approx. 1,500 vehicles per hour. This figure also applies to accessways that are greater than 4.2m wide[29].

Circular ramps of radii 7.5m, 9.0m and 12.0m measured to the outer kerb, with lane widths of 3.65m minimum have operational capacities of 1,200, 1,460, and 1,700 vehicles per hour respectively, but these figures may need to be reduced if the approach to enter or exit the ramp is interrupted.

Narrow ramps with poor visibility have reduced flow capacity as drivers exercise greater caution. Wide easy ramps with good sight lines and parking ramps allow drivers to travel more quickly, but if circulating vehicles travel at too great a speed it can increase the risk of accidents and injury to pedestrians.

In car parks with long search paths and where the drive aisles can become congested by drivers manoeuvring to and from the parking bays, it is advisable to introduce additional ramps to shortcut the circulation system and to allow drivers to move more quickly to the levels above. As a rule of thumb, single search paths or more than 500 bays benefit from having shortcuts. These additional ramps may be needed only on the lower levels of the car park where flows are greatest. However, in large car parks with tidal flow conditions and where the number of bays exceeds 900, the capacity of internal ramps may be the limiting factor in determining the dynamic capacity. In this case, D-shaped circular ramps or external straight ramps should be introduced where the number of bays does not exceed 1,300. Where the number of bays exceeds 1,500, circular ramps which are off the main circulation routes should be introduced — these are known as 'clearway ramps' (Section 5.4.7.3).

4.1.5 Bay turnover

Bay turnover (or 'churn') is the number of times a bay is used in a given period (usually a day) and is thus a measure of the car park use. It is calculated by dividing the number of cars entering the car park by the number of used parking bays. For instance, a fully occupied car park used solely by drivers parking all day would have a bay turnover of 1.0 and if the bays were not fully occupied the figure would be less than 1.0. Bay turnovers vary significantly and typical turnovers in excess of five cars per day have been reported for short-stay car parks which indicates considerable internal movement.

Attention is drawn to the recommended bay width of 2.7m for short-stay parking as opposed to 2.6m for medium- and long-stay parking. The increased width facilitates loading/unloading and helps drivers manoeuvre in and out of the bays. The relative ease of using bays where markings comprise parking pads or a double line with a 300mm space dividing the bays affords a better user experience.

In addition, for car parks with high peak flows or high bay turnover, short multiple search paths are desirable. These should be laid out systematically to help drivers search for an empty bay while minimising the number of vehicles circulating (Section 5.4.7).

4.1.6 Vehicle reservoir at entrance

In exceptional circumstances the peak hour arrival rate may exceed the capacity of the entrance barriers and/or the dynamic capacity of other parts of the car park, and vehicles may have to wait to enter. To prevent a queue extending onto the external highway, a vehicle reservoir should be provided before the entry barriers (Figure 4.4). The design of the reservoir should ideally allow a return to the highway without the need to enter the car park.

Figure 4.4: Vehicle reservoir at entry

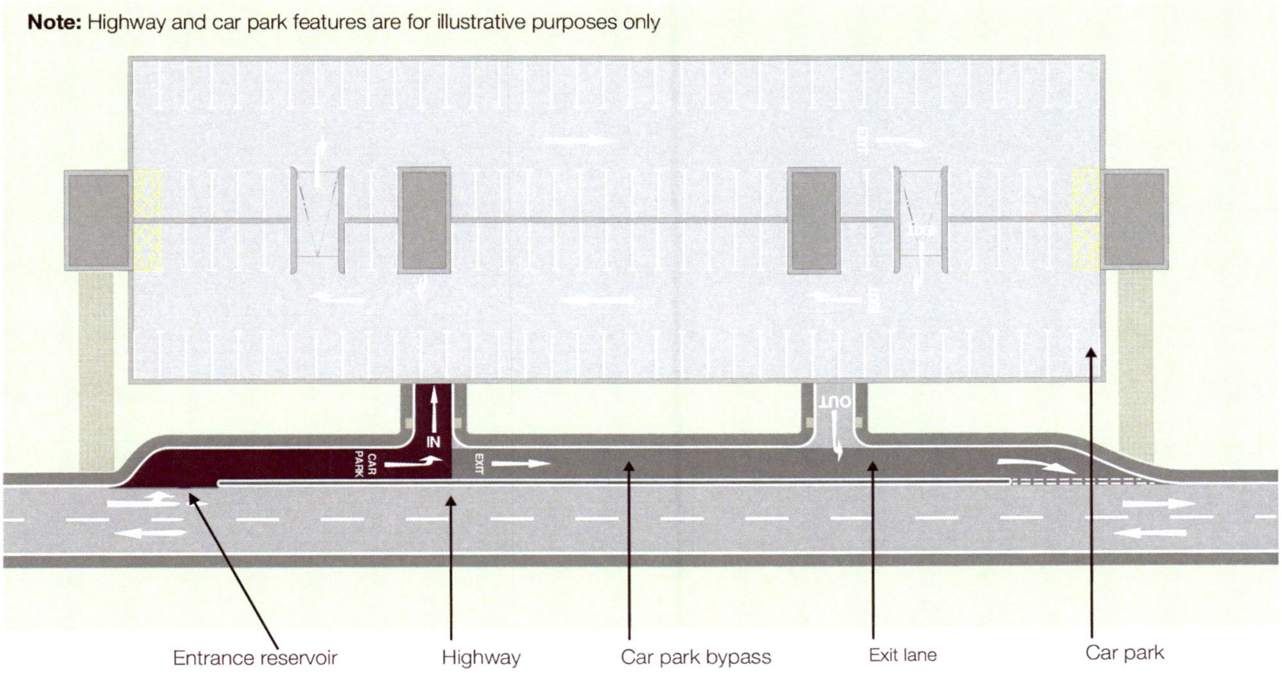

4.1.7 Vehicle reservoir at exit

If the exit barriers are to function at their maximum capacity, vehicles must be able to move away as soon as the barrier opens. If vehicles have to wait to enter the highway, a queue will form and obstruct the barrier and possibly the circulation inside the car park, resulting in congestion and an unacceptable reduction in dynamic capacity. It is therefore essential that the local authority is made aware of the peak exit flows to ensure that these can be accommodated by the highway and the associated junctions. If there is a shortfall, a vehicle reservoir will be required between the exit barriers and the highway; the design of which should minimise queuing at the exit barrier.

4.1.8 Design variations

The design parameters set out here are intended to provide general design guidance, and variations may occur due to the locality of the car park. While primarily for the design of UK car parks, this guidance can be applied to the design of other European car parks, with due account being taken of driving on the right-hand side of the road. Where the guidance is used outside the UK, local standards, bylaws, Building Regulations etc. should be considered, along with any special insurance requirements.

In other countries, alternative standards may apply (e.g., to cater for larger vehicles) and the design guidance will have to be reviewed and implemented accordingly. However, car designs are becoming increasingly focused on cars that are sold worldwide and so national and regional variations are less common.

4.2 Traffic management

For a car park to operate efficiently, traffic must be able to flow smoothly in, out and around the car park. Smooth entry and exit can be obtained by installing the appropriate number of entry and exit lanes to accommodate the anticipated peak flow and, if possible, some reserve capacity to allow for unintended consequences e.g., equipment malfunction.

In reality, of course, each time a vehicle manoeuvres into or out of a bay, or slows to search for a vacant bay, it has the potential to disrupt the flow within the car park. This can be reduced by using bay guidance systems which show, at high level, where vacant bays are located. Typically, the system will have a red/green light mounted above each bay (Figure 4.5), connected to a sensor that detects whether or not the bay is occupied. Sensors can be floor mounted or positioned above the bay looking down. Ceiling-mounted sensors are often preferred as they avoid the risk of being damaged by movement on the deck.

Figure 4.5: Car park bay guidance system

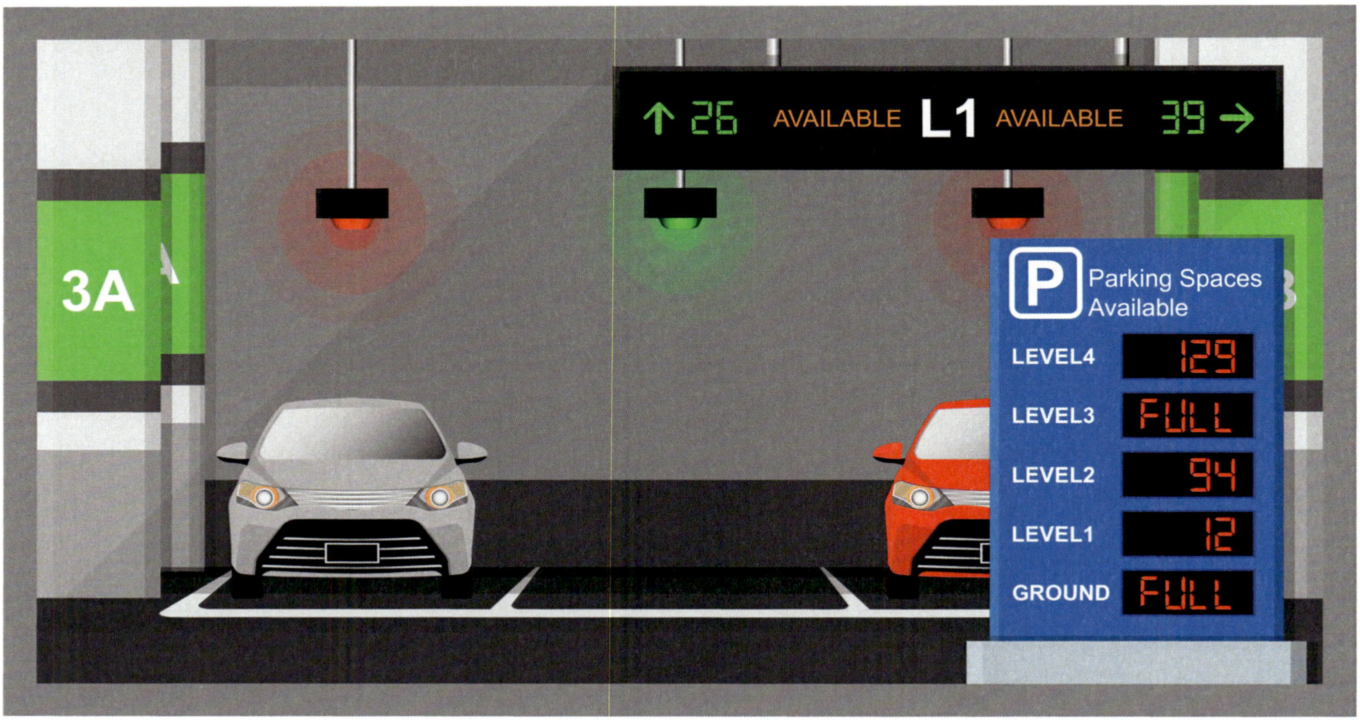

If the bay is empty a green light is displayed. If the sensors are well tuned, they will also detect when a vehicle is badly parked and obstructing the adjacent bay, showing the adjacent bay as unavailable. The systems of individual bay markers can be linked to a floor counter to show how many vacant bays are available on each floor. Some systems use a blue indicator to show where accessible bays are located. It is good practice to use symbols (e.g., a cross or tick) as well as colour to make this information accessible to people with red/green colour blindness.

In smaller car parks where the natural route through the car park would pass every bay, such systems are of limited value. However, they are invaluable in larger more complex sites, where the systems can allow drivers to bypass aisles or even floors where there are no vacant bays. Such a system will save driver time and reduces noise, pollution, and wear and tear on the structure.

4.2.1 Entry and exit

Access to and egress from a car park should be controlled for site security, revenue control purposes or to manage flow to and from the site. By far the most common method is to use a rising arm barrier (Figure 4.6). Barriers present a visible physical presence but are not designed to resist a vehicle impact. Such a barrier may also be linked to a variable sign to show whether there are spaces available. The barriers may also be linked to an Automatic Number Plate Recognition (ANPR) camera.

If the car park also provides spaces for cycles and motorcycles it is recommended that they have dedicated entry and exit routes, both for reasons of safety and to simplify the vehicle count. Motorcycles often enter and leave a car park by driving around the end of a barrier in the car lane which is less safe and can cause operational problems. It is recommended that cycles and motorcycles are parked as close to their entrance as possible in order to avoid them mixing with other vehicles on the ramps and aisles. This would also avoid potential problems with counting loops.

The entry/exit barrier will usually be linked to a system to identify the vehicle. This could be:

- To issue a manually-read ticket (read on exit) and used to let the vehicle pass or to calculate the fee due
- To issue a machine-readable ticket or token that will be used to calculate the fee due and be revalidated for use to open the barrier at an exit, often called 'pay on foot' (PoF) as the payment is made by the driver at a machine which is located away from the exit lane

Figure 4.6: Common entry/exit arrangement: rising arm barrier

- An ANPR system to identify the vehicle. ANPR may also be used with one of the other systems
- An on-vehicle radio-frequency identification (RFID) tag to identify the vehicle
- Where a car park has a pre-booking facility, the driver's arrival is usually detected using ANPR, or by entering the payment card used to reserve the space, into the entry gate card reader

The car park may also have open, barrier-free access. This could be because the operator does not wish to limit access to the car park, for example at a shop, or has reserved the site for specific users with a separate enforcement process. The car park may also accept payment via a pay and display system, with payment meters on the decks. Some car parks also use ANPR to identify vehicles and to allow calculation of a fee due, to be paid before exiting or to identify the vehicle's length of stay, if stay at the site is time-limited. With barrier-free access the risk of non-payment is significantly increased and so the operator will require a separate follow-up and payment enforcement system.

Manufacturers tend to describe the performance of their equipment in terms of the barrier cycle time i.e., the time from closed to open or open to closed and the read times of their ticket machines. However, in order to determine the peak flow that can be accommodated, the designer should provide a reasonable estimation of the number of vehicles able to pass through the lane in a given time period. Where there are multiple lanes merging or diverging before or after the barrier, the configuration needs careful consideration to ensure they operate without disruption.

For the car park to operate reliably at peak flow, it is desirable to provide enough lanes that even when one lane becomes unavailable, usually because of an equipment failure, the car park will still operate without queuing. If this is not possible, failures could be reduced by increasing the number of ticket machines.

The following figures are provided as guidance for the most common control systems:

- Manual ticket issue/fee calculation: 180–200 vehicles per hour (vph)
- An automated payment system where the driver takes a machine-readable ticket or token on entry, makes payment within the car park and presents the ticket to the exit lane ticket reader to exit the car park. A variation is where the driver can use a debit/credit card to "check in" and "check out" of the car park. Such systems would typically

allow passage of about 360vph. Sometimes such systems are linked to an ANPR camera system. This can be for security, to ensure that the ticket is being used for the correct vehicle, but it can also speed up the operation of the barrier by reading the vehicle number plate and opening the exit barrier without the need for the driver to enter the ticket. This facility could increase capacity by 50–75vph
- A system where the vehicle has a RFID tag or similar and the barrier opens when the tag is read, allowing the barrier cycle to be short-circuited if the following vehicle is identified before the barrier closes. This means that some 500 vehicles per hour can pass. Tags used as season tickets are sometimes mixed with the hourly parkers
- ANPR cameras can also be used in place of tickets, with or without barriers. With a barrier, a flow of 420–450vph should be achievable. Without a barrier, in a free-flow lane, flows of 500vph or more should be possible. However, it is important to ensure that adequate vehicle spacing is maintained to allow a clear view of the number plate

4.2.2 Rising step barriers and rising bollards

Rising step barriers and rising bollards (Figure 4.7) provide increased security at entry and exist because, unlike a rising arm barrier, the step/bollard provides a robust physical barrier that will either stop or badly damage a car that attempts to drive over it. Indeed, rugged versions are provided for site security which can even stop a heavy vehicle. Rising step barriers and rising bollards are often used in conjunction with a set of traffic lights to indicate whether they are in the up or down position.

However, the cycle time (the time taken to open and close after a vehicle passes) is greater than for a rising arm barrier. Also, if the rising step is struck by a car, it can become jammed. This may be desirable from a security viewpoint but renders the lane unusable until the barrier is repaired. Rising steps are not advised where a high traffic flow is anticipated.

4.2.3 Speed shutters

Speed shutters are full-height vertically operated shutters that protect the entry and exit lanes from entrance by pedestrians (Figure 4.8). As the name suggests, they open and close quickly and add to site security by excluding unauthorised access.

Figure 4.7: Rising bollards

Figure 4.8: Speed shutters

4.2.4 Speed control

Car park decks are a mixed environment of pedestrians and vehicles and so vehicle speeds need to be managed for pedestrian safety. A speed limit of 5mph (8kph) is recommended to minimise the risk of serious injury to pedestrians.

Generally, the distance that a vehicle can drive in a car park before reaching a point where flows merge or cross and the vehicle has to give way, is quite short and so the opportunity for vehicles to build up speed is limited. However, in some car parks there can be lengthy aisles. Where these are intersected by cross aisles, priority and permitted movements should be clearly shown with signs and deck markings, and careful consideration should be given to providing good sight lines, especially when the car park is busy and bay occupancy is high.

For long uninterrupted aisles, consideration should be given to the use of speed humps. However, the headroom for a vehicle driving over the hump would be reduced. Also, the impact on the deck can be quite large especially when a vehicle crosses the speed hump. Consideration might also be given to a lower (rumble strip-type) surface to raise driver awareness of the need to moderate their speed.

4.2.5 Pedestrians

Car occupants will possibly spend as much time in the car park on foot as driving. It is important to ensure the safety of pedestrians, and to provide guidance to help them to move safely and quickly to and from their vehicle.

On the deck itself, ideally the designer would provide a separated walkway for pedestrians (Figure 4.9). However, this may not be achievable due to either cost or site size considerations. Commonly, a pedestrian walkway is demarcated along the drive aisles on one or both sides of the driveway (Figure 4.10). This is not a reserved pedestrian way but prompts the driver to be aware of the presence of pedestrians. Pedestrians can and will cross the vehicle flow at any point to leave or access a car. However, where pedestrian flows are concentrated and cross the traffic stream, a pedestrian crossing-style marking should be used with appropriate signage. The use of signs that mimic the design and style of those used on the highway means that both drivers and pedestrians are familiar with their meaning and know how to react safely. There may also be a need for "traffic control" signs showing where pedestrians must or must not go, for example "No Pedestrian" signs on vehicle ramps. A driver who has possibly driven up (or down) several levels could become disoriented and therefore, wayfinding signage that correctly orients them is invaluable.

Figure 4.9: Segregated pedestrian walkway

Figure 4.10: Pedestrian walkway along drive aisles

Pedestrians will arrive on the parking deck via a lift or stairway. It is very important that these are located so that pedestrians do not walk out directly into the traffic flow. This is best achieved by creating a reserve area, which allows good intervisibility between pedestrians and cars, and where pedestrians can stand while they orient themselves on the deck. There must be step-free access to the deck to accommodate wheelchair users (who may not be using an accessible parking place), parents with children in pushchairs and anybody with impaired mobility or sight.

4.2.6 Closed-circuit television

Closed-circuit television (CCTV) can offer a number of benefits to the operator of a car park. It is essentially passive. Systems capture images and create a record; nothing more. Generally, they do not analyse the image. This requires human intervention and generally only occurs if there has been an incident.

The benefits of CCTV include:

- Continuous surveillance of the deck — allowing incidents such as crashes, attacks on pedestrians and vehicle break-ins to be recorded
- Allowing pedestrians using help points to be monitored
- General surveillance of entry and exit points and decks, assisting in the early detection of problems such as failed lighting units. Note: this is **not** a substitute for regularly walking the decks, stairwells etc. to ensure everything is in good order
- Added security for car park users — detection of loiterers etc.

"Smart" CCTV systems can analyse the movement of people within a car park and flag up abnormal behaviour. They can also detect fires at a very early stage.

CCTV systems are distinct from ANPR systems (Section 4.2.1).

4.2.7 Customer help points

Customer help points are generally situated in the stair cores and allow customers to make contact with the control room should they need urgent assistance. They should also be situated (along with helpline signage) in areas where there is a risk of self-harm or suicide e.g., on the open top deck of a car park.

4.3 Lifts and stair cores

4.3.1 Lifts

Requirements for lifts and escalators will vary significantly between different developments, but in multi-storey car parks will usually be required to satisfy legislation for equal access and to meet the reasonable requirements of users, e.g., those carrying shopping, children etc. This is closely linked to the internal planning and management of pedestrians and the capacity considerations outlined earlier in this chapter. The use of the car park (e.g., short/medium/long-term applications) will have a significant impact on the occupancy which should be used for the purpose of any lift traffic analysis, although it should be noted that the provision of lifts will, in many car parks, be dictated by minimum legislative provisions rather than by meeting lift traffic performance standards.

General guidance in relation to performance benchmarks and calculation guidance is provided in CIBSE's Guide D[30].

Early consideration of the firefighting and evacuation strategy (in conjunction with any appointed fire engineer) will also be key to determining the requirements for either firefighting or evacuation lifts. In the UK, the requirements for firefighting shafts and the need for these to incorporate firefighting lifts are provided in BS 9999[31]. This is further discussed in Chapter 14.

In the context of internal planning, it is worth noting that goods and service lifts should not be sited within firefighting shafts, although where required, firefighting and evacuation lifts can double as passenger lifts for standard operation.

Further information on lifts can be found in Chapter 13.

4.3.2 Stair cores

The stairs cores within a car park can provide both day to day access for users and emergency access for firefighters.

The number of stair cores and the level of fire protection and firefighting facilities will be determined by the building's fire strategy. Further advice on the fire aspects of stairs is covered in Chapter 14.

Once the technical requirements of the stair cores are covered, consideration should be given to the user experience, particularly for stairs anticipated to be in daily use (rather than reserved for emergency access only).

This can include the design of wall, floor and soffit finishes, the daylighting/lighting strategy, the design of balustrades and handrails incorporating jump deterrents to any voids (Chapter 12), the type of doors used within the stairs and any lobbies, and the visibility of the stairs both from inside and outside the building.

The stairs should provide an attractive and welcoming space to encourage users to select the 'active travel' option and promote good health, while ensuring that fully accessible lifts are also equally visible and available to users. This means the stairs should be easy to locate within the building, ideally next to the lifts, and users should be able to see into the stair core before entering it in order to improve the perception of safety and security[17].

Where possible, clear glazed windows should be provided into the stair core from outside. This can also provide an opportunity for natural smoke ventilation at each level (if required by the fire strategy) rather than reliance on an enclosed smoke shaft. The windows offer improved natural surveillance, helping to deter antisocial behaviour within the stair cores (e.g., urination or drug taking), and can provide daylight to the space, reducing the reliance on artificial lighting.

The finishes within the stair core should be selected with regard to the intended user group/s and maintenance/ design life requirements of the client. Floor finishes can range from exposed concrete, to a resin waterproofing finish, to porcelain/ceramic tiles. Tiling may also be applied to walls to half height (e.g., handrail level), or even full height in situations where an extremely robust finish is required. Consideration must be given to achieving sufficient visual contrast between floor, wall and handrail, and possibly to the provision of anti-graffiti coatings. All floor finishes must provide adequate slip resistance in line with the Building Regulations[10]. Accessibility requirements for stairs are covered in Chapter 3.

Barrier matting (robust "carpet-like" continuous matting set in a recessed well to create a flush finish with the floor) is recommended at ground floor level for stair and lift cores in daily use. This helps to pick up water being tracked in on shoes in wet weather, which can create a slip hazard even on slip-resistant flooring if not mopped up promptly.

Soffit treatments may range from exposed precast concrete, to a painted finish, to a suspended ceiling. As cores are typically unheated spaces, appropriate products must be chosen (e.g., an external grade ceiling panel or a mineral paint product which will not blister and peel). Where suspended ceilings are used, they must be anti-tamper and robust to prevent vandalism and the risk of people hiding objects (typically weapons or drugs) within the ceiling void.

Signage should be provided within the stair core itself, as well as externally, to identify floors and help people locate the floor they have parked on, before they leave the stair. This is also an essential aid for the fire service.

5 Design geometry and layout

5.1 Introduction

To design an efficient, high quality, safe and easy to use car park it is essential to evaluate all the constraints and to arrange them in order without allowing one to dominate to the detriment of the car park as a whole. Achieving the best compromise may entail challenging the constraints as part of the overall design process, but this will lead to a better outcome.

The priority is to ensure that the final design best suits the intended purpose and above all its users, as they can often spend significant time in the car park over many return visits. Large capacity car parks with a high turnover of bays need to be dynamically efficient with good visibility and therefore the internal circulation pattern requires careful selection. Each circulation pattern has limited capability in terms of speed and ease of use. Small, low turnover car parks are less dependent on the circulation pattern and compromises can be made more easily.

The external highway network and the position of the entry/exit into the car park significantly affects the choice of circulation pattern and the ramping system for vehicles to move vertically. Designing the internal exit arrangement and controls to handle the predicted number of vehicles will result in failure if, for example, multiple entry/exit lanes are unable to merge efficiently and the highway network has insufficient capacity to handle the peak flows from the car park.

It is important to note that the average life of a motor vehicle is approx. ten years whereas the design service life of a car park is much longer. Throughout this period the building has to be able to accommodate changes in vehicle sizes, weights, means of propulsion and even self-parking systems. There has to be sufficient flexibility in the initial design to enable modifications to be made without detriment to the appearance or functionality of the car park. The types of vehicles that currently enter a car park can vary from large sports utility vehicles (SUVs) to sports cars, wheelchair accessible vehicles (WAVs), emergency service vehicles, motorcycles, cycles and electric scooters.

The range and variety of car designs is such that it is never going to be practical to design a car park that will accommodate every possible variation — from the lowest, sleekest supercar and cars with gullwing doors to the largest pickup truck. This guidance cannot encompass all vehicles, but aims to cater for 95% of the vehicles that make up the current fleet on UK roads.

Entry and exit controls are commonly used in both private and public car parks; not only to restrict use to those authorised and to collect payments but also to prevent entry when the car park is full. While technology changes rapidly over time, it is nevertheless important to ensure that adequate provision is made in the design for expected peak entry/exit flows, otherwise it may be difficult to operate the car park satisfactorily at a later stage.

Accessible links are needed to connect the car park to external facilities. The design of footways, crossings, standing areas adjacent to lifts and doorways, and routes from accessible parking areas should be reasonably level, have good visibility and suitable clearances to enable all user groups to move safely throughout the car park.

Many factors influence whether a user will find the car park easy to use and feel comfortable within the facility. The most significant elements are:

- Size/capacity of car park and ease of circulation
- Presence or absence of internal structure and if so, column layout
- Ability to easily find available bays
- Aisle and ramp widths, headroom and ramp gradients
- Accessibility
- Ventilation and air quality
- Safety and security

- Lighting and level of visibility
- Quality and style of internal surface finishes
- Clear and concise user information, wayfinding and signage

Dimensions in this chapter are net, and allowances should be made for finishes and protrusions into the parking areas. Where parking bays and walking areas are demarcated by lines on floors, the dimensions are to the centre of the lines. Special consideration and different standards will apply if the car park is required to provide access routes for large emergency vehicles.

5.2 Vehicles

5.2.1 Cars and light vans

A review of manufacturers' details for new vehicles available in the UK[32] has identified a significant change in vehicle dimensions in recent years. The major change has been the progression from 4x4s and multi-purpose vehicles (MPVs) to sports utility vehicles (SUVs) and the introduction of hybrid and battery electric vehicles (BEVs).

For simplification when using this guidance, the range of vehicles has been grouped and a 'standard design vehicle' (SDV) has been introduced to represent 95% of the vehicle fleet found on UK roads.

The examination of new cars available in the UK shows that the height of 95% of standard cars fall below 2m, exclusive of roof racks/boxes. However, adding a roof box increases the SDV height to approx. 2.5m.

The range of typical vehicle dimensions is given in Table 5.1.

The turning circle is the minimum space required for the outermost part of a vehicle to undertake a complete U-turn — and can range from 10.2m for a typical small car to 12.5m for the SDV, to around 13.5m for large vehicles. Some of the large vehicles are fitted with rear wheel steer which reduces their turning circle. These are minimum kerb-to-kerb turning circles and include overhangs of the car body and the start of the turn before achieving full lock. All these factors contribute to creating the path that is generated by the turning vehicle (known as the 'swept path').

A simple template of the SDV and its swept path is shown in Figure 5.1.

5.2.2 Wheelchair accessible vehicles (WAVs)

WAVs are those that are specially adapted for wheelchair users and may be taller than the recommended clear internal height (Section 5.3.8) but are no heavier than 3,000kg and require special provision (Section 5.3.3).

5.2.3 Motorcycles

Motorcycle parking is often required, either by local planning policy (e.g., a Parking Standards Supplementary Planning Document, or a parking clause within the local plan), or by the car park client. However, wherever possible, efforts must be made to avoid motorcycles sharing access with cars. In particular, the internal ramps present a hazard to riders, and motorcycle parking should therefore be located as close to the entrance and exit points as possible in

Table 5.1: Typical vehicle dimensions

Vehicle group[a]	Length (m)	Width[b] (m)	Height[c] (m)
Small vehicle[d]	≤3.95	≤1.75	≤1.75
Standard design vehicle[d]	4.90	2.00	2.00
Large vehicle[e]	≥4.95	≥2.05	≥2.05

Notes:
[a]Includes electric and hybrid vehicles.
[b]Excluding wing mirrors.
[c]Excluding roof boxes, racks and roof bars.
[d]Includes 95% of vehicles on UK roads.
[e]Larger than 95% of vehicles on UK roads.

Figure 5.1: Swept path of standard design vehicle

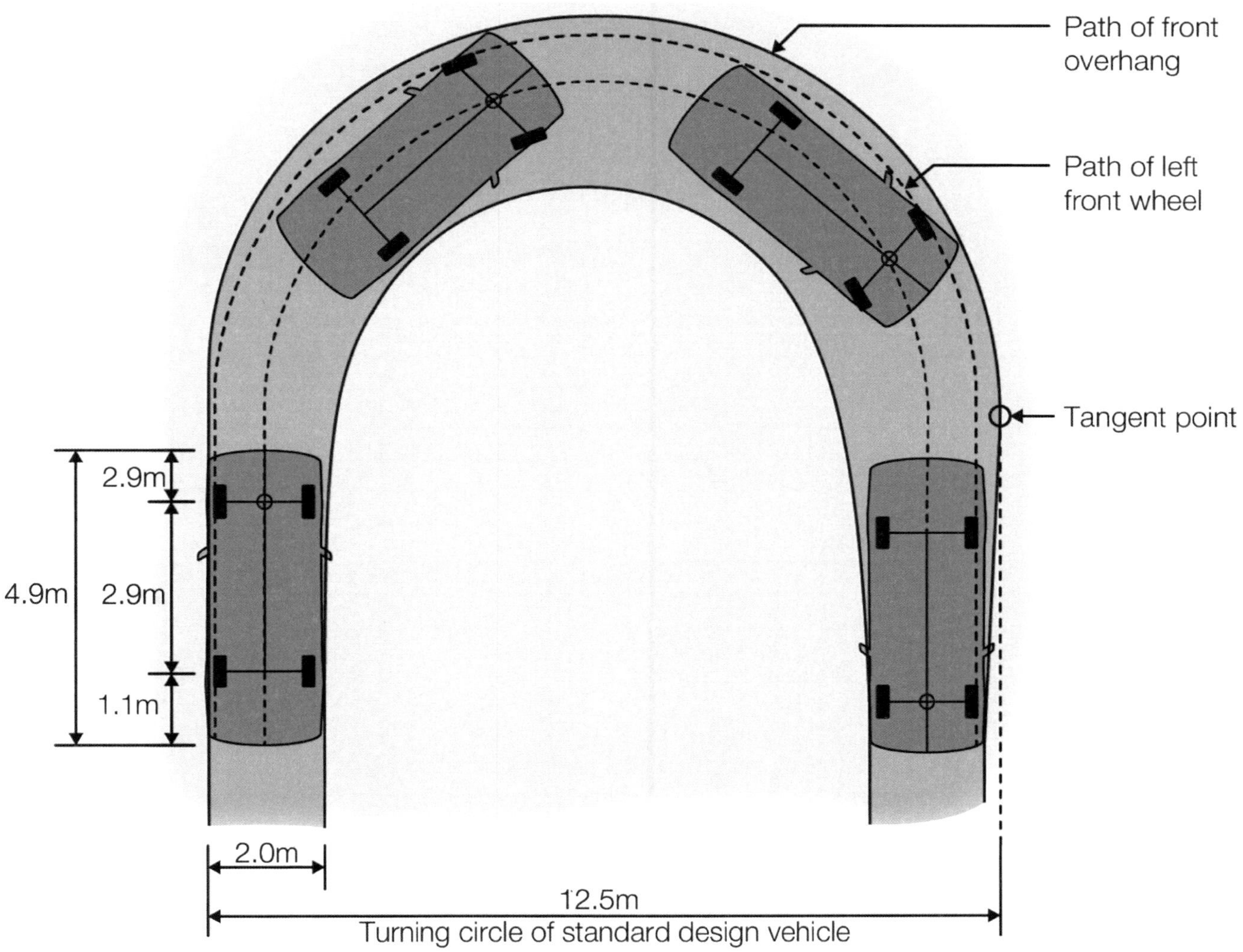

order to minimise the need to circulate within the car park. Motorcyclists are often provided with free parking and therefore need the option to bypass barrier systems, which can be dangerous for riders. Pay and display is not viable for motorcycles, but ticketless systems have been implemented by some local authorities who wish to charge for parking.

The ideal scenario is to provide a self-contained motorcycle parking area with separate access for cars. Provision should be made for a safe pedestrian route to and from the primary access point with minimal crossover of car routes. For security and user perception of safety, the motorcycle parking area should have excellent visibility and be clearly visible to staff (where present on site), other users or passers-by. Motorcycle theft is a significant concern for riders. Natural surveillance deters criminal activity and helps users feel more comfortable. The motorcycle parking area should also be covered by good lighting levels and CCTV, in line with the strategy for the rest of the building.

Motorcycle parking areas can either be identified as individual bays or as a larger shared area, usually demarcated by line painting. The latter can promote more efficient use of the space and is usually preferred. The parking area must be flat (or have a minimal fall to aid drainage) as motorcycles may fall over if parked on a slope of 5° or greater. The surface must be able to support the weight of the motorcycle and the point load applied through a prop stand or centre stand, which can be considerable. Bitumen based surfaces may not be suitable as they can become soft during hot weather and deform under the weight of the stand.

Where individual bays are identified, a minimum bay size of 1.4m x 2.1m is recommended as this can accommodate the majority of motorcycles, which vary significantly in size from small mopeds through to large cruisers.

Provision must be made for a locking rail or individual locking hoops or posts to provide a robust point for riders to secure their machines to, using their own chains or locks. Flush-mounted locking rings set in the floor are generally discouraged due to the trip hazard they present, and the difficulty in cleaning and maintenance. Proprietary systems are available, but an alternative can be a bespoke steel rail along the back of the parking area. This should be positioned at approx. 600mm above the surface level to accommodate a range of wheel sizes.

In a private car park, it may be appropriate to provide locker storage for helmets and other rider equipment. This should be located close to the motorcycle parking area but consideration must be given to the building's fire strategy, as storage areas can present a fire risk on escape routes etc.

Electric motorcycles are gradually increasing in popularity and are typically charged using a proprietary cable while the battery is in place on the motorcycle. Consideration may therefore be given to providing some charging points, or futureproofing (with ducting and data/power supplies) for these to be fitted in the future to accommodate demand.

5.2.4 Cycles and electric scooters

Increasingly, there is an expectation that cycle parking will be provided. Cycling is growing rapidly across the UK, particularly in urban areas, and cycle users needing to visit the location for the same reason as car drivers (e.g., work or leisure) also require parking. This includes a significant uptake in the use of cargo cycles and non-standard cycles.

When considering the provision of cycle parking, the designer should consider the following:

- How many cycle parking spaces are required? This figure may be provided by the client or be defined in local planning policy (including the London Plan for developments within the Greater London area, which represents one of the more ambitious standards nationally)
- What level of security is required? Staff cycle parking may be fully secure with access via key fob or swipe card, whereas public short-stay parking may be open to all. Subscription models for commuter cycle parking are increasingly popular, with access control via key fob or code providing access only to those who pay a regular fee
- Are any other facilities required? These could range from lockers, cycle maintenance equipment and battery charging points for E-bikes, through to showers, changing rooms and drying rooms. Note that accessible facilities must be provided in accordance with Building Regulations to cater for all users. As these facilities grow in complexity, they have an operational and maintenance demand which must be considered by the client. Third-party operators manage a number of subscription-type facilities for public use, typically at train stations or city centre locations and will take responsibility for operation and maintenance in exchange for revenue generated from subscription fees

Cycle parking (whether private or public) should be located in close proximity to the main pedestrian access to the car park. Separate access to cars is essential for safety and convenience, but consideration should be given to minimising conflict with pedestrians. Cyclists should not need to cross pavement areas to get to or from the cycle storage.

As with motorcycles, cycle theft is a major issue. Cycle storage areas should therefore be located in highly visible positions with good natural surveillance. Appropriate lighting and CCTV coverage is also of critical importance.

Cycle storage areas should incorporate a mix of rack types in order to provide a range of options for different cycle types and user abilities. Non-standard cycles can include handcycles, tricycles, cargo bikes, recumbents and even the size of a 'standard' bicycle can vary from a narrow lightweight road machine to a fat-tyred off-road cycle with broad handlebars.

Racks that only support the front wheel must not be used as these are not secure and can damage the cycle. Sheffield stands (Figure 5.2) are regarded as the most accessible type of rack, with appropriate spacing (typically a minimum of 1m between stands). However, as these are not the most spatially-efficient option, the designer may wish to also include a proportion of tiered racks or semi-vertical racks (Figure 5.3a and b) which can fit more cycles into a smaller area. Tiered racks will require additional headroom to enable use of the upper rack. This varies by manufacturer but is typically 2.7m. For safety and ease of use, all types of racks should be positioned on a level surface.

Figure 5.2: Sheffield stand

Figure 5.3a: Tiered cycle racking

Figure 5.3b: Semi-vertical cycle racking

Cycle storage areas should be located on the entrance level of the building wherever possible. If located on lower or upper levels, dedicated ramps and lifts need to be provided. Ramps must be of an appropriate gradient and width depending on whether users are to be encouraged to ride or push their cycles, and lifts should be dual entry to avoid needing to turn or reverse the cycle out.

Private e-scooters are not currently legal for use on public land in the UK, but an increasing number of trials for hire scooters are ongoing in cities around the country with multiple providers — who are responsible for the operation, maintenance and charging of their vehicles. The e-scooters are either required to be returned to specific docking locations, which have been agreed with the relevant local authority, or can be parked anywhere on public land. Geofencing can be used to control where the e-scooters can be taken or parked, and is an increasingly important tool to prevent pavement 'clutter' and hazardous abandonment e.g., blocking pavements or crossings.

If the government legalise the use of private e-scooters in the future, car park designers and clients should consider whether they wish to provide dedicated parking or charging facilities. E-scooter users will potentially require similar facilities to cyclists; including lockers, drying rooms and changing facilities. These could therefore be shared between different 'active travel' user groups.

If a public-sector client wishes to make provision for e-scooter parking or charging, coordination will be required with the third-party provider over their specific requirements.

There is a potential fire risk associated with e-scooter batteries, which has led to Transport for London (TfL) currently banning their use or transport on TfL services. Consideration should therefore be given to the location of any e-scooter area within the car park and its impact on the building's fire strategy.

5.3 Geometry

5.3.1 Introduction

Parking layouts should be configured to allow drivers to manoeuvre easily and safely, and to segregate pedestrian areas and routes from vehicular routes. The ease of manoeuvring is a function of aisle and bay widths and can significantly influence the dynamic capacity. This is particularly important where a high turnover of bays is required e.g., car parks that serve convenience shopping facilities, where aisle, ramp and entry/exit capacities are critical to the functionality. For car parks that serve facilities where customers are familiar with the parking arrangements, for example an office car park, it may be possible to modify the guidance, providing it does not adversely affect the performance of the car park and render it unfit for its intended purpose.

UK geometric requirements may differ from those of other countries.

5.3.2 Bay width and length

Current practice is to design the car park for use by a wide range of vehicle types; collectively represented by the 'standard design vehicle' (SDV) which encompasses 95% of the most commonly used vehicles. Larger cars can be accommodated, provided that the height does not exceed any restrictions, but their ease of manoeuvrability may be diminished, especially during busy periods.

The anticipated use of the car park and the need to accommodate EV charging equipment influences the size of the parking bays. The facilities served by the car park generally determine the anticipated use. However, as there can be many variations and the use may vary throughout the day, it would be misleading to base the bay dimensions solely on the nature of the facility. Therefore, for simplicity and for the purpose of identifying the anticipated use of the car park, the duration of stay (Section 4.1.2) has been chosen as the means of selecting the bay dimensions (Table 5.2).

Parking bays for medium- and long-stay parking are classified as 'standard' parking bays and those for short-stay parking and comfort parking (Section 5.3.4) are classified as 'enhanced' parking bays. The dimensions of the standard parking bay have been 2.4m x 4.8m for more than 60 years. Today's SDV is approx. 200mm wider, with doors that have become progressively thicker — reducing the space needed for vehicle occupants to enter and leave the vehicle comfortably, and without risking damage to adjacently parked vehicles. While still suitable for small cars, the standard bay width is now considered too small to accommodate today's SDV.

In any parking arrangement there is usually a mix of large and small cars — and an average has been taken in determining the dimensions of parking bays and the need to house equipment for charging EVs (Table 5.2). Wherever possible, consideration should be given to harmonising the services and equipment in order to accommodate them within the proposed range of parking bay dimensions.

Increasing the width of parking bays has the benefit of improving dynamic efficiency as drivers can park more rapidly by having to make fewer turns to enter bays that are at 90° to the drive aisles. It also improves sightlines when leaving the parking bay.

If it is necessary to provide larger parking bays than those shown in Table 5.2 they should be within a dedicated area. Smaller parking bays can also be provided and marked 'small car bays' within a dedicated area but there is a risk of disruption from larger vehicles trying to park in them.

Where there is a need to increase the density of parking for vehicle storage e.g., hotels and airports, the vehicles are parked in very close proximity in blocks. Each block is divided into rows of minimum width to allow the vehicles to be parked sequentially. Members of the public are not allowed to access these areas and they are classified as 'special structures' (Chapter 20).

Table 5.2: Parking bay dimensions

Type of parking	Length[a,b] (m)	Width[a] (m)	Comment
Short-stay	5.00	2.70	High turnover of bays
Medium-stay[c]	5.00	2.60	Mixed use
Long-stay[c]	5.00	2.60	Low turnover of bays
Accessible	5.00	3.80[d]	–
Parent/child	5.00	3.40[e]	–

Notes:
[a]All dimensions are to be clear of any projections (Section 5.3.6).
[b]The preferred dimension is 5m for all bay lengths. However, with restricted space and appropriate signage, this can sometimes be reduced for small/city vehicles[17].
[c]Bays for medium- and long-stay parking are classified as 'standard parking bays'
[d]The bay width for use by disabled persons allows for the door to be fully opened to improve movement in and out of the car and to provide greater room for assistance to be given to those with reduced mobility. An additional 1.2m wide cross-hatched area is required to the driver's side and to the rear of the vehicle[24,25].
[e]The bay width for use by parent and child allows for the door to be opened more fully for access to child seats.

5.3.3 Accessible parking areas

These bays will be sized in accordance with regulations and will therefore include an additional zone on both sides and to the end of the bay alongside the drive aisle. This generally provides enough space for most disabled users to access their vehicle including equipment such as a wheelchair or mobility scooter†.

BS 8300[25, 28] recommends that a proportion of accessible parking bays are provided with a clear height of 2.6m, which may also be a requirement if the car park serves a particular facility. The additional clear height facilitates the use of a rooftop wheelchair box, or most high-top vans (where the user may be sitting on a secured wheelchair rather than a car seat within the back of the vehicle). While not particularly common, the wheelchair rooftop box is designed to assist manual wheelchair users who can transfer from the vehicle into the wheelchair without assistance but cannot load it into the boot of the vehicle, thus facilitating independent travel for these users. In a multi-storey car park, it may be possible to provide a zone with 2.6m clearance for accessible bays at ground level, rather than increasing the height of all floors. It is important to ensure that the vehicular access route maintains this height for the entirety of the route and clear signage is provided to indicate any change in height constraints within the building. Where it is not possible to accommodate these taller vehicles, consideration should be given to providing adequate facilities outside the car park or nearby on-street.

Accessible parking bays should be located in close proximity to lifts and the primary circulation core of the building in order to minimise horizontal travel distance. Positioning all accessible bays at ground floor may not be the most appropriate solution if it significantly increases the travel distance for users. Where accessible bays are distributed across multiple floors, they should be positioned in a consistent location for ease of wayfinding. The designer should also consider positioning the bays away from the busiest circulation routes, in order to give more time for users to manoeuvre and to take equipment in and out of the vehicle.

For more detailed design requirements for accessible bays, refer to Chapter 15.

5.3.4 Aisle width and bin width

Guidance for aisle and bin†† widths for various parking angles with bays on each side of the aisle is shown in Table 5.3. These preferred dimensions should be clear of any structure or edge details (Sections 5.3.6 and 5.3.7). Aisle widths are designed to accommodate a walkway and to allow the SDV to manoeuvre into and out of the parking bays. Where demarcated pedestrian paths are shown within the aisle, it is not as a reserved space, but to raise driver awareness of pedestrian traffic.

† The additional 1,200mm space should be in addition to the recommended bay size.
†† A 'bin' is two bays situated either side of a drive aisle. If the bays are 5m long and the drive aisle is 6m wide, the bin width will be 16m (5m + 6m + 5m).

Table 5.3: Drive aisle and bin widths

Parking angle	Preferred aisle width (m)	Bay width (m)	Bin width for 5m bay length (m)
90°	Two-way aisle: 6.95	All	16.95
90°	One-way aisle: 6.00	All	16.00
60°	4.84[a] 4.75[a] 4.74[a]	2.50 2.60[c] 2.70	16.00 16.01 16.10
45°	4.26[a] 3.75[a,b] 3.60[a,b]	2.50 2.60[c] 2.70	14.86 14.50[b] 14.49[b]

Notes:
[a]Minimum widths required for the SDV to manoeuvre allowing 500mm clearance from other parked vehicles when making the turn from the drive aisle into the parking bay.
[b]These dimensions do not take into account the need to provide a 1.2m wide pedestrian walkway within the drive aisle. Ideally the aisle should be increased in width to 4m with a corresponding increase in bin width.
[c]Standard parking bay

To overcome constraints imposed by limited space or particular user operations, variations to these dimensions can be considered, but when this results in reduced dimensions the client should be made aware of the variations and the resulting limitations e.g., restricted space between parked vehicles and more difficult manoeuvring.

Where more generous or comfort parking is specified, an increase in the preferred dimensions may be necessary, but consideration should be given to whether it can be accommodated by standard structural systems without the need to introduce additional columns internally, as these would adversely affect the ease of use and dynamic efficiency of the car park. A possible alternative is to configure the columns so they do not project into the parking bays — as this will allow the bays to be remarked should vehicle sizes change over time.

Oblique parking (where the parking bays are angled at less than 90° from the drive aisle) is dynamically more efficient and certainly more pleasant to use, but it is not as efficient spatially. Even when considering the reduced width of the drive aisle, oblique parking still requires more net floor area per parking bay. Also, the minimum aisle width necessary to accommodate vehicles entering and leaving a parking bay should be treated with caution as the drive aisles are shared spaces and therefore sufficient width is required to accommodate safe pedestrian routes.

5.3.5 Deck and pedestrian ramp gradients

A minimum fall of 1:60 is required to facilitate drainage of the decks and to prevent ponding.

Where the decks are used for access to lifts and stairs, other facilities, and to and from accessible parking areas, the gradient should be no more than a standard drainage fall of 1:60. Delineated walkways are usually located within the width of the drive aisle and are best positioned on one side only. They should be a minimum width of 1.2m, with markings to indicate that they are for pedestrian use.

Where decks are sloped to form a parking ramp, the gradient should not exceed 1:20. If the gradient is steeper, difficulties will be experienced in opening and in keeping open a car door on the up-gradient side and in closing a door on the down-gradient side. Also, unbraked shopping trolleys, pushchairs etc., can roll away when unattended. Accessible parking areas should not be situated on parking ramps and should be located in areas that are almost flat (Chapter 15).

Flatter gradients and minimum cross falls are preferred in motorcycle and cycle parking areas (Sections 5.2.3 and 5.2.4).

5.3.6 Clearance to the structure

For regular structural grids, the end parking bays and those that are situated adjacent to the ramps or stair cores tend to be narrower which makes these bays less suitable for SDVs and more suitable for use by smaller vehicles only. A similar situation exists with shear walls and large columns that project into the parking bays. Consideration should be given to increasing the width of these parking bays by around 300mm.

For split-level decks on narrow sites, it is possible to overlap the decks in order to reduce the overall width of the car park, but the projections from the overhanging decks can potentially damage vehicles while being parked (Figure 5.4).

Figure 5.4: Overlapped decks

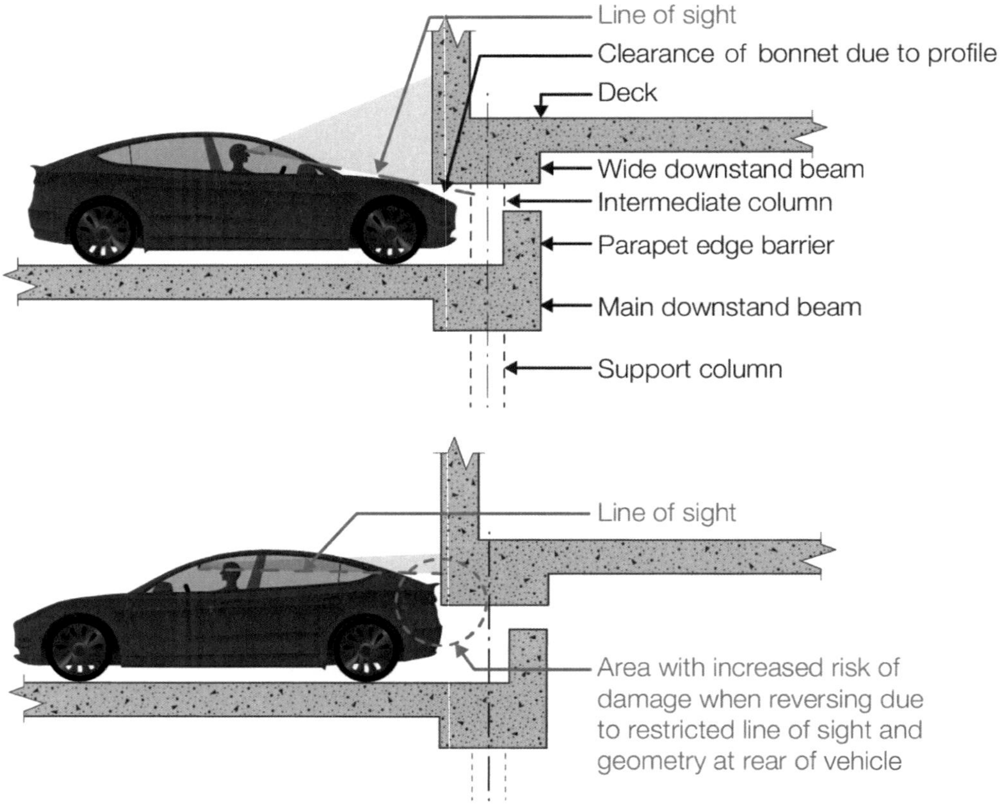

5.3.7 Column location

Clear-span construction is preferred, as it provides a safer environment for both drivers and pedestrians, but in mixed-use car parks and in basements other considerations can dictate the need for internal columns. The sizing and spacing of the columns need careful consideration in order to maintain efficient bay access and good sightlines. Columns at the front of the bays i.e., near to the drive aisles will reduce accessibility, and in order to improve parking manoeuvres the columns should be set back from the drive aisles (Figure 5.5). Columns within the middle third of the bay can obstruct car doors and should be avoided.

A column projection of 150–200mm into the parking bay is acceptable, provided that columns are within the recommended setback zone from the aisle (Fig. 5.5). Where larger columns are required e.g., in mixed-use facilities, special attention is needed to maintain satisfactory clearance and manoeuvrability. Ideally, the building and car park grids need to be coordinated to ensure that the column grid is not detrimental to the functionality of the car park.

For the purposes of futureproofing, consideration should be given to locating the columns outside of the parking bays and incorporating them into the cladding — as this will free the bays of all columns and allow them to be relined to suit future changes in vehicle size.

5.3.8 Headroom

The minimum headroom, i.e., clear height measured normal to the deck or ramp to the structure or any projection for signs and services etc, should not be less than 2.2m. This has been established to suit the SDV, and speciality cars (e.g., with gullwing doors) may have difficulty in opening the doors, due either to the proximity of the adjacent parked vehicle or the headroom inside the car park.

The minimum headroom applies also to the entrance and exit areas, so careful attention needs to be given to the provision of services and operational requirements for each of these areas (Section 5.3.11).

Figure 5.5: Column positions

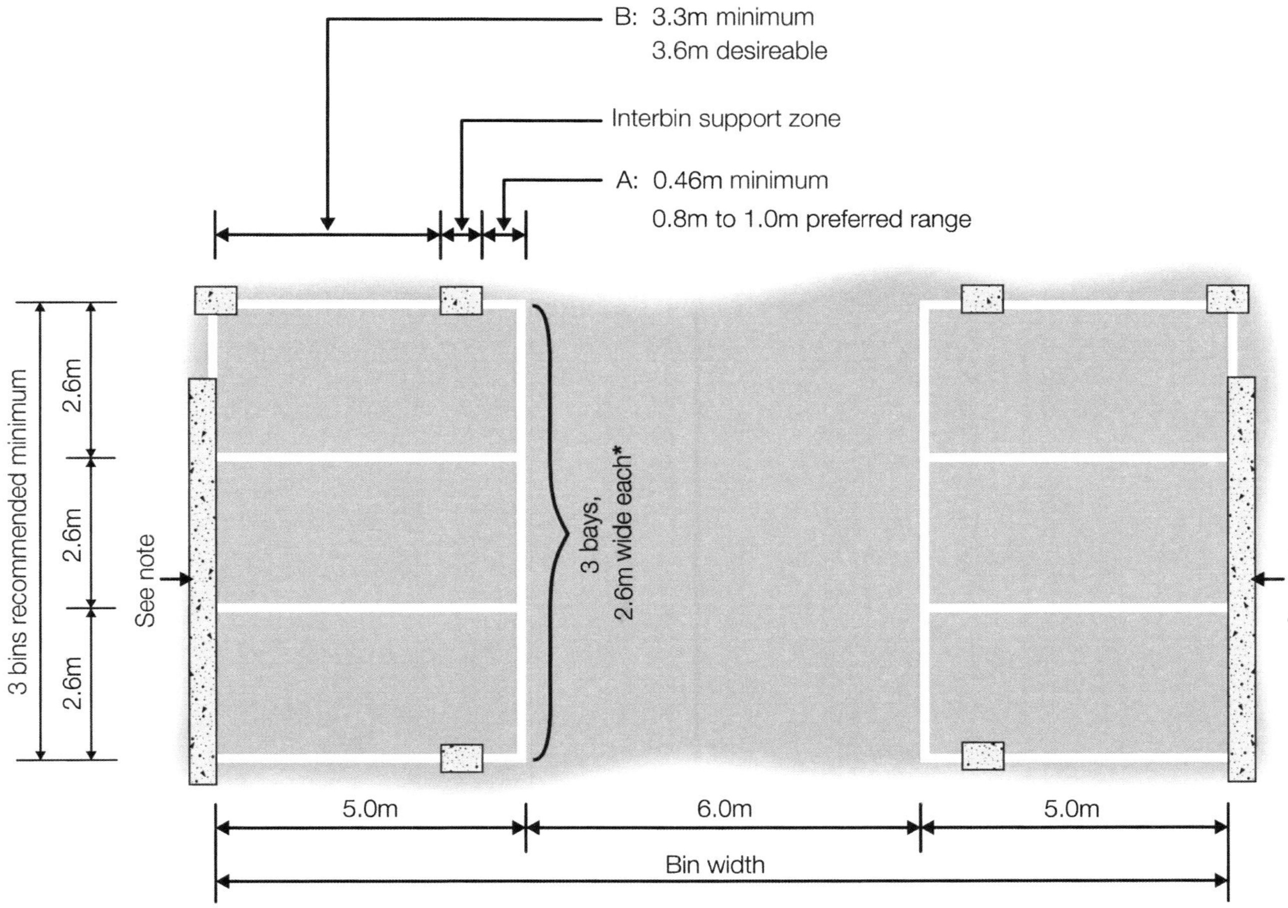

Note: For linear supports (e.g., impact resistant walls), where columns are required, they must be positioned solely on the interface between the bays.

Key: Acceptable support positions * Typical bay dimensions

Where the car park is large on plan, increasing the minimum headroom will enhance the internal environment, but consideration must be given to the ramp geometry to ensure that the gradients or the projection of the ramp into the drive aisle does not become excessive. Increasing the headroom may also be applicable to car parks where a greater proportion of vehicles with roof boxes is likely e.g., in tourist hotspots.

Additional clearances are required at changes in gradient e.g., at entry/exit zones to ramps, as the effective height of the vehicle increases (Sections 5.3.9 and 5.3.10). Traffic-calming measures, such as speed humps and tables, must be carefully located as they are typically 75–100mm high and can restrict headroom locally. Rising-step traffic control barriers (Section 4.2.2) require pits approx. 300–600mm deep which should be taken into account when considering the available headroom on the floors below.

For safety, the headroom indicator board at the entrance to the car park is normally set some 50–100mm below the actual headroom within the car park. This reduction should be discussed and clarified with the eventual operator. If the height restriction is imposed by low service ducts and/or pipes, the indicator board needs to be supplemented with a height restriction positioned so that the critical component will not be damaged by an over-height vehicle.

To determine structural height, outline designs should be prepared for signage, lighting, ventilation, barrier controls, sprinkler systems, conduits and drainage pipes and any other possible projections below the structure. The downward projections of the various services should be estimated plus an allowance for finishes, dimensional tolerances and structural deflections, and together added to the headroom to determine the clear structural height required.

5.3.9 Ground clearance

The ground clearance of the SDV is 100mm, which is measured from the deck to the lowest point of the body of the vehicle. It is particularly important when considering the design and location of speed humps and the transitions at sudden changes in floor level (Figure 5.6).

Vehicles with ground clearances less than 100mm are considered 'special vehicles' and grounding could be an issue.

Figure 5.6: Effect of changes in floor level to ground and overhead clearance

5.3.10 Vehicle ramp requirements

Ramps are the means of moving vertically throughout the car park and are where many minor scrapes take place as a result of their width being inadequately proportioned for current vehicles. Their size and gradient significantly affect the dynamic efficiency of the car park, as vehicles have to reduce speed to change direction when negotiating the ramps. Ramp orientation is a fundamental aspect of the design layout (Section 5.4.7) and the minimum geometric requirements are outlined in the following three subsections.

5.3.10.1 Straight ramps

Long, straight, one-way ramps require the minimum width between kerbs to be not less than 3.2m, but when it is necessary to turn onto or from the ramp e.g., from the drive aisle, the entrance and exit ends of the ramp should be widened (or 'flared'). The amount of necessary flare depends on the ramp width and the manoeuvre required to enter or leave the ramp. Clear sightlines are essential in these locations.

Where one-way ramps are short and the ends are not flared e.g., between split-levels, a minimum constant width between kerbs of 3.65m should be maintained. Any structure to form the edge of the ramp should be set back a minimum of 300mm from the face of the kerb (Figure 5.7). Where kerbs are omitted, any structure should be set back a minimum of 300mm from the theoretical edge of the kerb, and the margin marked with line marking and cross-hatching (Table 5.4). Taking into account the thickness of any structure/barriers to the sides of the ramps, the overall width of the ramp should fit between two standard parking bays.

On two-way ramps, a central margin of minimum width 0.5m is required and should be formed in line marking and cross-hatching without the use of kerbs. Structure that could impair sightlines on the ramps should be avoided, but where it is necessary to have a column within the central margin, the minimum side clearance to the column should be 0.3m (Fig. 5.7). While two-way ramps can be accommodated between three standard parking bays, they are not

Figure 5.7: Straight ramp widths and clearances

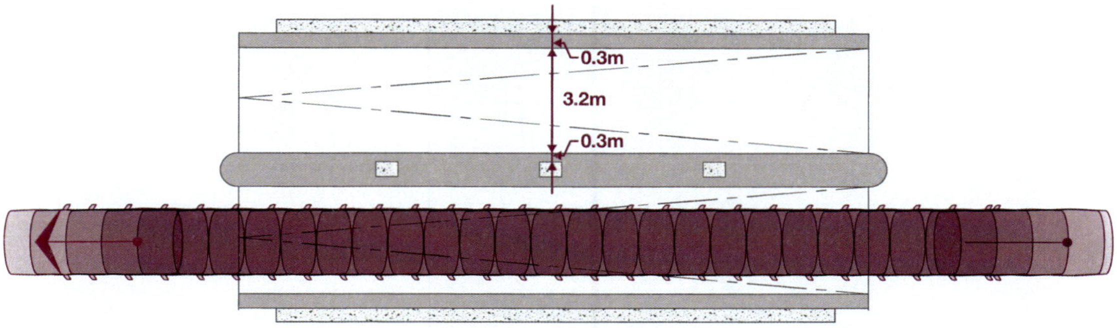

Ramp with straight entry/exit

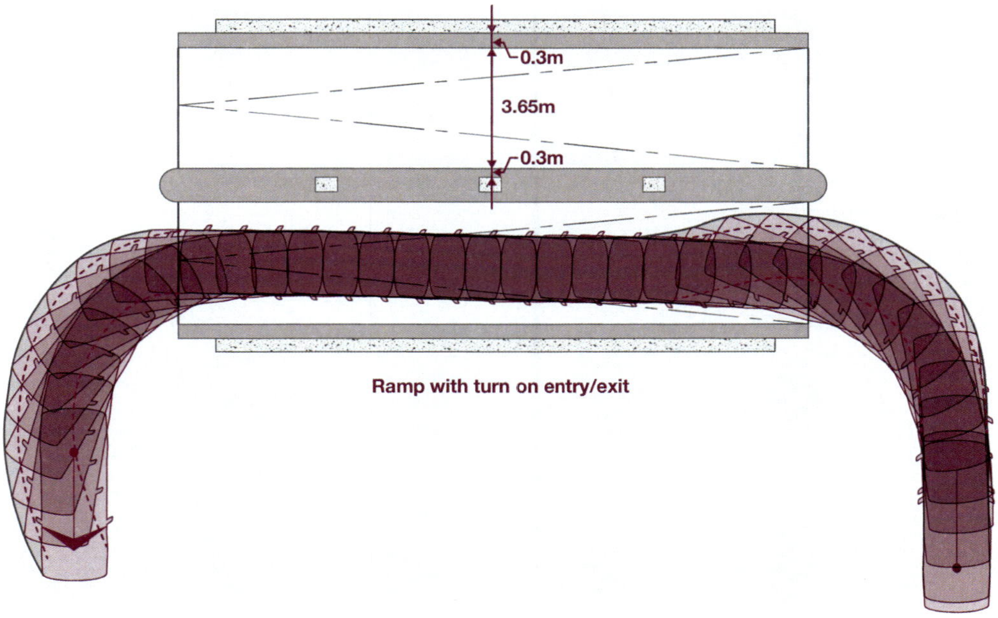

Ramp with turn on entry/exit

dynamically efficient owing to their being relatively narrow which causes vehicles entering or leaving the ramps to slow down or stop. Therefore, for large capacity car parks and car parks with a high bay turnover, two-way ramps that fit between four parking bays are preferred.

The maximum gradients for vehicle ramps should not exceed those given in Table 5.5. However, if site constraints dictate that steeper ramps are required, extreme caution should be exercised, and a design risk assessment prepared to ensure that the gradients and transitions have no unresolved operational risks.

Table 5.4: Minimum widths between kerbs (straight ramps)

Type of ramp	Position	Width (m)	Additional side clearance to structure (m)
One-way	Width for straight approach	3.20	0.30
	Entry/exit section for turning approach	3.65	
Two-way	Width for straight approach	6.90[a]	
	Entry/exit section for turning approach	7.80[a]	

Note:
[a]Width includes a column- and structure-free central margin of 0.5m. Where central margin contains columns or structure refer to Fig. 5.7.

Table 5.5: Maximum gradients for vehicle ramps

Type of ramp	Rise (m)	Maximum gradient
Straight	Not greater than 1.5	1:6[a]
	Greater than 1.5 and less than 3.0	[a, b]
	Greater than 3.0	1:10
Curved	Not greater than 3.0	1:10[c]
	Greater than 3.0	1:12[c]

Notes:
[a]With transition slopes top and bottom.
[b]Maximum gradient reduces linearly for ramp rises of 1.5–3.0m.
[c]Gradient measured on centreline.

If vehicle ramps are steeper than 1:10 or the floor is laid to a fall of 1:60 or greater, away from the top of the ramp, a transition length is required. These transitions should be sited at the top and bottom of the ramps to reduce the risk of vehicle grounding. The transition length should be at least 1.8m and the gradient should be half the gradient of the ramp (Figure 5.8). The transition can extend into the drive aisle but the difference in level between the drive aisle and the edge of the transition (taken at the interface between the drive aisle and the parking bays) should not exceed 50mm otherwise it may be difficult to merge the different levels without forming a trip hazard. Detailed sketches of transitions for ramps are given in *Car Park Designers' Handbook*[29].

5.3.10.2 Curved ramps

The turning circle of the SDV is 12.5m but for a large car it can be approx. 13.5m. When a vehicle begins its turn, it sweeps out a wider path, and when combined with the overhangs of the body the actual diameter is greater than the turning circle (Section 5.2.1 and Fig. 5.1).

Curved ramps (including circular and D-shaped ramps) change in direction along their length. The preferred radius between outer kerbs for one-way curved ramps is 9.0m. and the absolute minimum should not be less than 7.5m (Table 5.6).

The minimum width of the ramp between kerbs should be 3.65m and clearance to the structure should be added (which in the direction of travel is 0.6m to the outside kerb and 0.3m to the inside kerb).

Figure 5.8: Typical ramp section

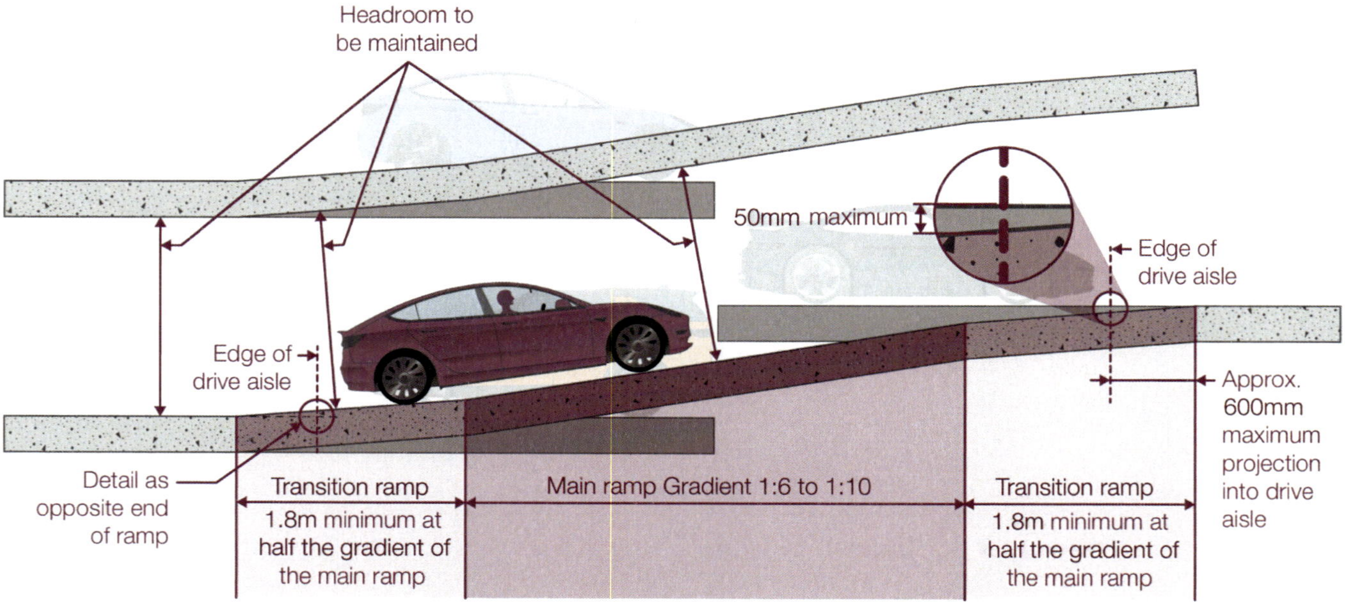

Table 5.6: Minimum outer kerb radius (one-way curved ramps)

Option	Radius (m)	Structure clearance to outside kerb (m)	Structure clearance to inside kerb (m)
Recommended	12.00	0.60	0.30
Preferred minimum	9.00		
Absolute minimum	7.50		

For two-way curved ramps (Figure 5.9) the minimum width between kerbs should be 7.8m, which incorporates a central margin that is 0.5m wide and formed in line marking without the use of kerbs (Table 5.7).

5.3.10.3 Superelevation

Curved ramps benefit from having superelevation which should be a maximum of 1:20. However, care must be taken to ensure that the cross fall and gradient along the inside kerb is not too steep — so that inclined vehicles are at a sufficiant distance from any structure on the inside of the ramp.

5.3.11 Entry and exit arrangement

To prevent queuing at the point of entry, the capacity should be equal to or greater than the estimated maximum peak hourly flow. A vehicle reservoir is required between the highway and the entrance barriers to prevent vehicles from

Figure 5.9: Preferred minimum diameter to kerbs (two-way curved ramps)

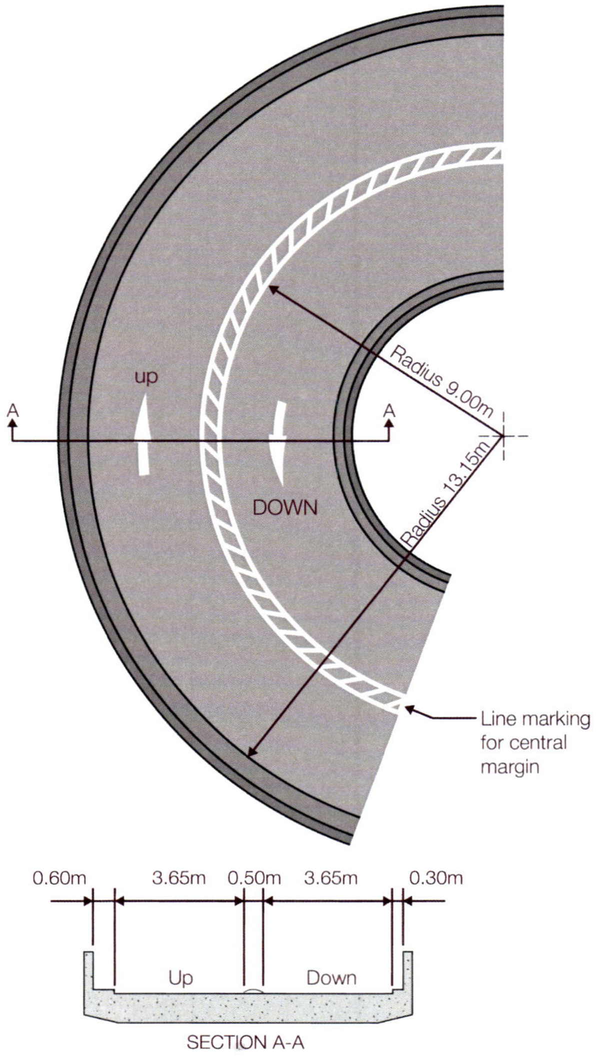

Table 5.7: Minimum width between kerbs (curved ramps)

Type of ramp	Width (m)	Width central margin (m)	Structure clearance outside kerb/margin (m)	Structure clearance inside kerb/margin (m)
One-way	3.65	n/a	0.60	0.30
Two-way	7.80[a]	0.50		
Note: [a]Width includes a column- and structure-free central margin of 0.50m (Fig. 5.9).				

queuing on the highway and to provide a transition zone to reduce speed from the highway before entering the car park. If multiple entry lanes are merging, there must be a vehicle reservoir/weaving area of sufficient capacity to allow vehicles to merge and enter the car park without blocking the entry lanes.

It may sometimes be appropriate to provide a suitable turning area to avoid entering the car park and passing the entry barriers. If this cannot be accommodated, the parking equipment should be configured to allow drivers to enter and exit the car park (e.g., within ten minutes) without being charged.

The external highway and junction capacity will both influence the rate of flow from the car park. It is not satisfactory if the highway network cannot cope with the peak flows from the car park as this can lead to internal congestion and user frustration. Therefore, this potential issue should be raised with the local authority to see if a satisfactory solution can be found and whether a vehicle reservoir is required. Careful consideration should also be given to the merging of lanes after the exit barriers to ensure that the space is sufficient to avoid congestion before entering the highway. Layout of a typical combined entrance/exit is shown in Figure 5.10.

It is often advantageous to position the entrance and exit side by side. The number of lanes is determined by dividing the peak hour flow by the operational capacity of the chosen parking equipment (Section 4.1.3). It is good practice to round the required number of lanes upwards to allow for equipment or vehicle malfunction. Where space is limited and this is not possible, one or more lanes can be designed as reversable; then, if peak inbound and outbound flows occur at different times, a lane or lanes can be reversed as appropriate. This will reduce cost, add flexibility and minimise the risk of blockages should equipment fail.

Where it is only possible to accommodate one entry and one exit lane, the potential disruption from equipment failure can be minimised by providing duplicate parking machines working in tandem.

If the lane width (either adjacent to the parking equipment or at a pay station) is too wide, drivers can experience difficulties in not being able to reach the ticket machines or kiosk. This reduces the dynamic capacity as drivers have to lean out of the windows, open doors, or even get out of their vehicles to use the machines. Hence, drivers are encouraged to approach ticket machines and pay stations as closely as possible by restricting the width of the lanes to 2.5m (Fig. 5.10).

Positioning of parking equipment on a bend is not advisable as drivers have difficulty in reaching the ticket machines. Where the approach lane is curved, it should be widened to ensure that vehicles can stop in close proximity to the equipment. If space allows, a straight of at least 6m should be provided on the approach to the parking equipment. The area alongside the equipment should have shallow gradients to reduce the risk of vehicles rolling forward or backwards. Gradients of 1:20–1:40 are acceptable.

An emergency/late pay bay should be provided before the exit barriers to allow for a vehicle malfunction while close to the barrier or for a driver who has forgotten to validate their parking ticket to park temporarily.

5.3.12 Kerbs

Raised kerbs within the circulation areas should be omitted wherever possible as they can be a trip hazard. If used, kerbs need careful consideration with regard to the fixing detail and the interface with the deck waterproofing, in order to avoid maintenance issues.

Figure 5.10: Typical combined entry/exit layout

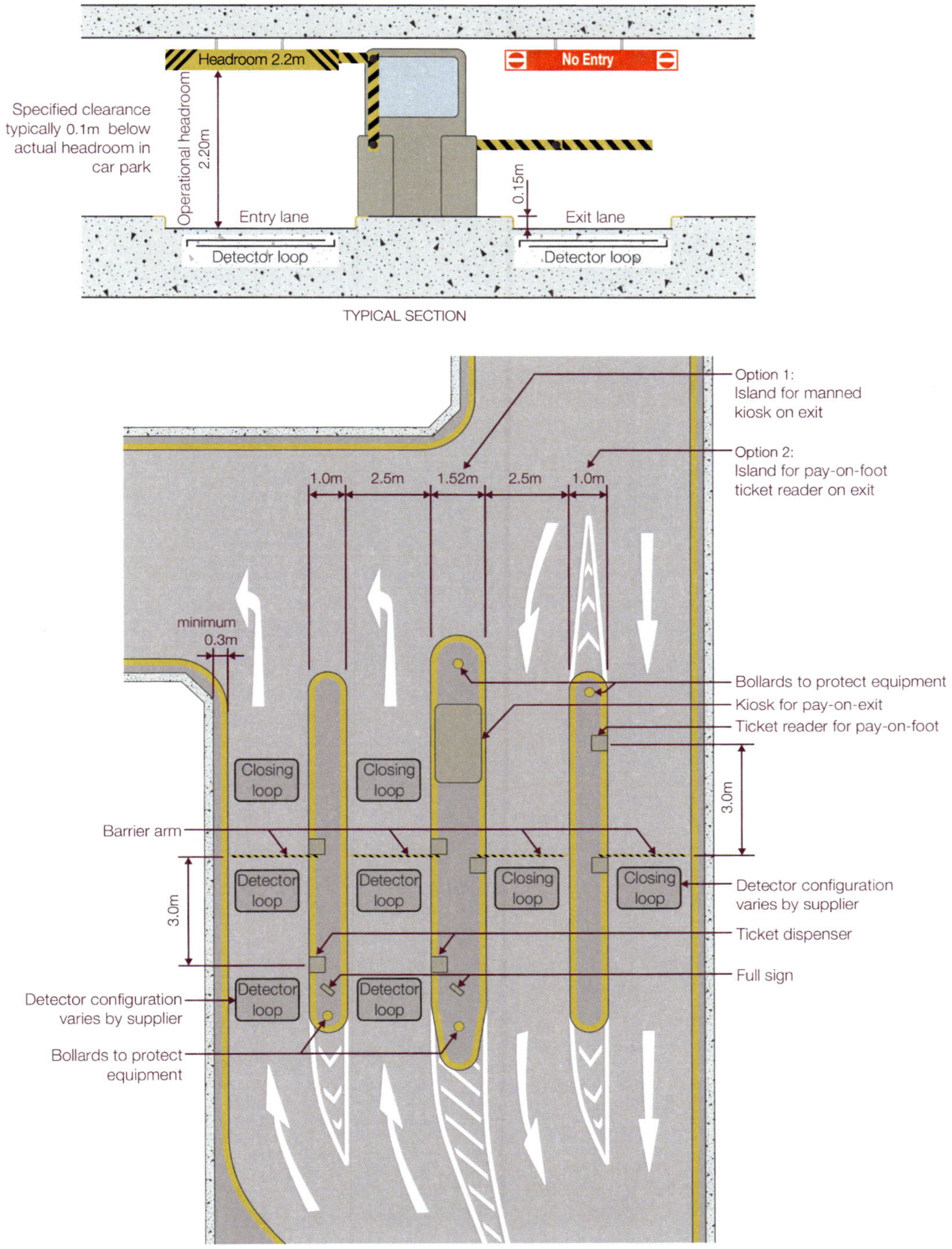

While raised kerbs within ramps will provide additional protection to the edge structure and details, they can increase the risk of vehicle damage. The use of kerbs to separate opposite flows on two-way ramps should be avoided as drivers on the down-ramp can lose sight of the kerb. Also, raised kerbs can unnecessarily restrict the movement of vehicles, and painted margins or rumble strips should be used in preference, provided that the edge structure has been designed to resist impact from an errant vehicle.

The recommended kerb height is 100mm and in areas that are required to be accessible dropped kerbs should be installed. Where it is important that cars do not mount the kerb, such as near control equipment, the minimum kerb height above channel level is 150mm. Such high kerbs can inevitably cause damage to wheels and in these areas rubber kerbs are preferred to concrete kerbs.

5.4 Layout

5.4.1 Design principles

Car parks have to be affordable, dynamically efficient, safe and easy to use for both the driver and all passengers. There are numerous options for configuring the internal layout and circulation, each having advantages and disadvantages, so determining the preferred option can be challenging. It is preferable to design a car park firstly from the inside, and to address the external appearance after the functionality has been optimised.

The design has to carefully consider how the car park will be used and how this may change over time. It must also be compatible with the locality and follow the guidelines established by the LPA in terms of appearance and scale.

Short-stay parking has a greater bay turnover for a given level of occupancy and so attracts more two-way traffic throughout the day. Medium-stay parking will have a mix of tidal flow and bay turnover, whereas long-stay parking produces high single-direction traffic flows with only a steady bay turnover.

The layout and internal circulation should therefore be chosen to suit the estimated peak hourly flows and the anticipated bay turnover. Car parks expected to have a significant bay turnover should preferably have one-way flow systems, as this will ensure the greatest dynamic capacity.

Drivers will often prefer to park on the level that is as near as possible to their final destination. Where this is close to the entrance it can quickly create congestion. The dynamic capacity is reduced when localised congestion results in near-empty parking levels that are at a greater distance from the final destination. A similar situation can develop at the exit particularly if the external highway is unable to cope with the number of vehicles leaving the car park. Careful planning of internal ramps and exit locations can overcome potential congestion issues.

Where the peak hour flows and bay turnover are expected to be high, provision should be made to short-circuit the lower levels by having a rapid up-route and, if necessary, a rapid down-route. Ideally, single search paths for incoming drivers should not exceed more than 500 bays (Section 5.4.7) — above which shortcuts should be introduced in order to maintain dynamic efficiency.

The maximum occupancy will generally be less than the static capacity owing to a number of factors. Not all drivers will park in the first vacant bay, parking discipline may be poor, and those already in the car park may miss vacated bays. Therefore, deliberate under-capacity margins of approx. 5% are sometimes introduced where the entry system is controlled.

It is essential be able to recirculate the various levels especially without having to exit the car park should a vacant space be missed.

5.4.2 Cul-de-sac parking

As bays within cul-de-sacs (Figure 5.11) are generally off the search path they are less busy areas. This makes them suitable for accessible parking or parent and child parking areas, provided that easy access is available to the lifts and stairs. The maximum length of a cul-de-sac should be limited to six bins of parking, but where site constraints make longer cul-de-sacs unavoidable, a vehicle turning space should be incorporated into the layout. Although it is preferable to increase the width of the end bays to facilitate ease of parking these bays should not be used as a turning area, as manoeuvring at the very end of the cul-de-sac is more difficult.

The direction of flow in cul-de-sacs is two-way so when incorporated into a one-way circulation system, it can be quite tight for a mix of passing vehicles and pedestrians to share a 6.0m wide drive aisle — hence the need to restrict the preferred length of the cul-de-sac to six bins.

5.4.3 Drive aisles

Drivers need to be able to find their way around the car park easily in order to concentrate on looking for a vacant parking bay. A one-way drive aisle circulation has a higher dynamic capacity than two-way systems and can reduce confusion and congestion. Drivers sometimes drive in the wrong direction which can cause disruption to the standard flow and potentially be a safety hazard. This can be managed by good floor markings and clear signage.

Figure 5.11: Cul-de-sac parking with wider end bays

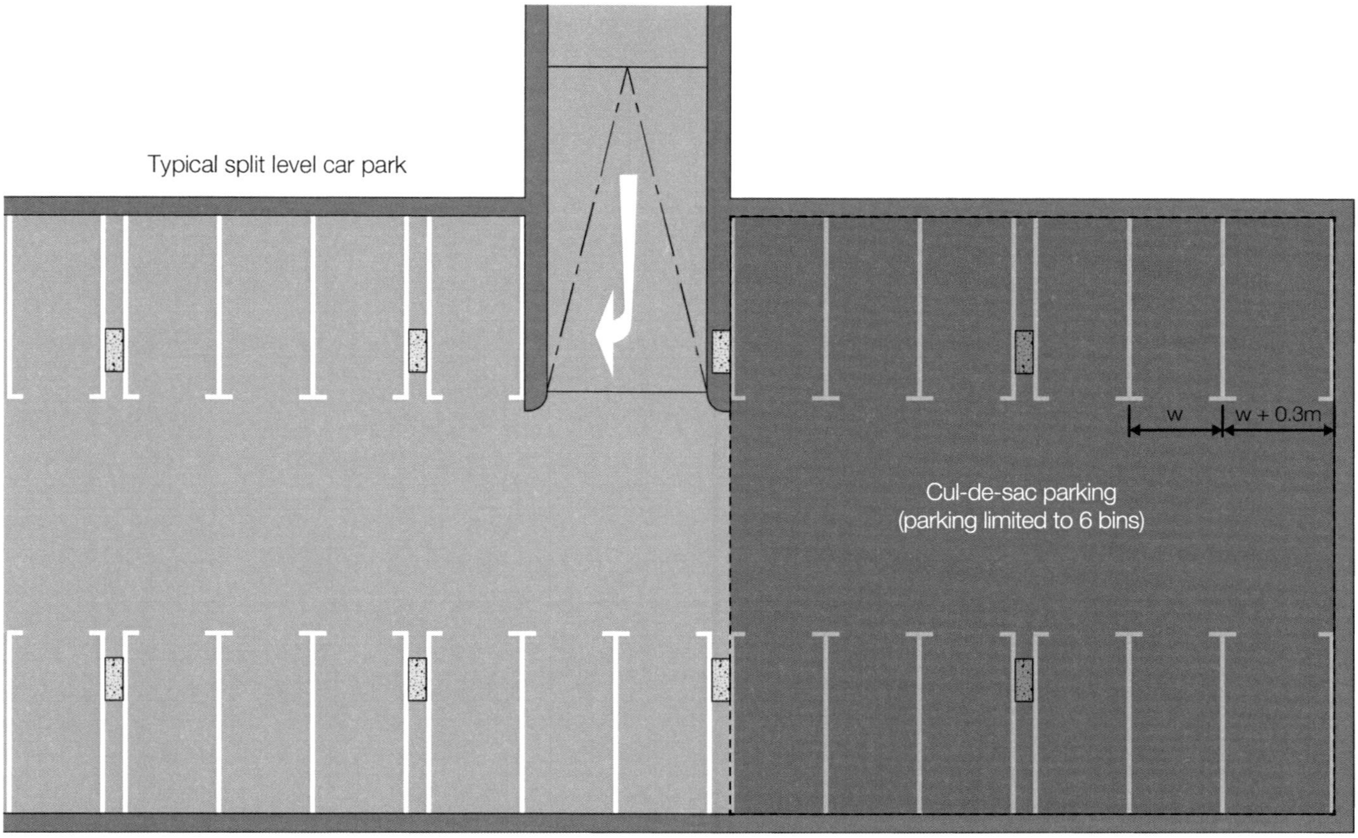

However, if the circulation pattern requires two-way drive aisles, the aisles should be wider with better visibility, good floor markings and clear signage. Cross-movements to enter or leave bays interrupt the traffic flow, making two-way circulation less dynamically efficient than one-way systems.

In larger car parks, long circulation aisles are inevitable and can lead to excessive vehicle speeds if controls such as speed humps are not provided. However, when specifying speed humps the following should be considered: pedestrian accessibility; restriction of headroom; deck drainage; impact loading on the slab and potential damage to the deck waterproofing; and the need to maintain the speed humps' fixing bolts. Rumble strips may therefore be preferable to speed humps.

5.4.4 Vehicle ramps

Ramps are required to enable vehicles to move vertically between the different parking levels. They can be straight or curved, short or long, lengths of inclined decks (parking ramps), and one-way or two-way. Irrespective of type, the ramp (or ramping system) has a significant influence on the dynamic capacity. This is due to the need to reduce speed in order to change direction or to give way to vehicles that have priority on the drive aisles — or in the case of parking ramps, to give way to vehicles that are making parking manoeuvres.

Additional ramps can be incorporated into the design to allow vehicles to move more quickly between levels without the need to circulate the main parking areas, or where a parking ramp has insufficient dynamic capacity.

In large car parks with high entry or exit flows or where there is a need for a high turnover of bays, one-way curved ramps that are circular will enhance dynamic efficiency — especially if they are positioned external to the car park and exclude the circulation routes. This type of ramp is known as a 'clearway' ramp. Two-way circular ramps are the least efficient dynamically as one of the flows needs to cross the other when entering or exiting the ramp, resulting in congestion inside the car park and a reduction in the dynamic capacity. For two-way circular ramps it is always better to have the inner flow as the down-route and the outer flow as the up-route as this is the shallowest gradient. An alternative to the two-way circular ramp is the 'interlaced' ramp where the ramps are interwoven on plan[29].

Where visibility is restricted by an inner wall, circular one-way ramps should flow in an anticlockwise direction with the driver on the outside of the turn, as this will give the greatest line of vision. It is also best suited to ramps that have the minimum radius to the outside kerb (Table 5.6). For larger ramps where anticlockwise flow is not possible careful consideration should be given to having a skeletal structure in place of the inner wall in order to maximise the sightlines.

Vehicle ramps are generally unpopular as there is an increased risk of accidents largely due to lack of visibility. As such, they need significant design input in order to reduce the risk (Section 5.3.10). Narrow ramps and scissor ramps are unpopular and are the least efficient dynamically. Scissor ramps in particular pose the greatest risk, as the flow on one of the ramps has to give way to the other, often in a situation of restricted visibility. This can be improved slightly by separating the ramps with a couple of parking bays, but even so, the use of scissor ramps (Figure 5.12) should be treated with caution.

In some layouts, traffic contra-flow occurs on straight ramps or crossovers between ramps. In these unusual cases, it is recommended that traffic streams are separated by an impact-resistant barrier or bollards which are clearly visible, as motorists may be required to drive on the opposite side or direction to the standard traffic flow.

Figure 5.12: Scissor ramp

5.4.5 Circulation

In the UK, motorists drive on the left-hand side of the road, and in a clockwise direction around roundabouts — so it seems natural when leaving a parking bay to set off in a clockwise direction. Therefore, the preferred direction of circulation within a UK car park is clockwise. Sometimes anticlockwise flow has to be adopted especially when it is governed by the entry/exit position in relation to the direction of flow on the external highway. In such cases careful consideration is needed with regard to the configuration of road markings and signage.

Wherever possible the entry and exit flows should be separated as this will lead to the greatest dynamic efficiency. Integrated flows where the exit route is within the main circulation route (Section 5.4.7.10) should only be used in very small car parks with a low turnover of bays.

The preferred minimum diameter in which to turn a vehicle through 180° is 18.0m (Table 5.6) and therefore it is difficult to turn from one drive aisle to another when the aisles are separated by a single row of standard parking bays. Two rows of parking bays are therefore preferred between drive aisles to enable vehicles to turn from one aisle to another.

5.4.6 Parking angle

Although creating bays at an angle of less than 90° is advantageous for drivers as they facilitate single turn entry and exit which in turn improves the dynamic capacity of the car park, angled parking is not widely used in the UK, especially in parking structures. However, when the site is too narrow to accommodate parking bays at 90° to the drive aisles, angled parking may be suitable. That said, angled parking is a much less efficient use of space. Parking angled at 60° uses about 15% more space than parking at 90°, and at 45° angled parking uses about 30% more space.

Although the drive aisles can be narrower with angled parking, care should be exercised to ensure that the narrower drive aisle has sufficient width to incorporate a pedestrian walkway (Table 5.3).

An increasing number of cars are now equipped with self-parking systems. These systems use external sensors to identify and measure potential parking spaces and then "drive" the car into a suitable slot. This isn't new; the first systems became available in the early 1990s, initially for parallel parking but more recently for parking bays in car parks. This creates a difficulty for angled parking as the systems are configured to back cars into bays. As self-parking systems become more commonplace the utility of such bays becomes questionable. Drivers with this functionality would prefer to drive in the wrong direction down the aisles, whereas drivers in cars without the equipment tend to drive in the correct direction and park nose first. This could of course be solved by providing a two-way drive aisle, but that would eliminate the rationale of using angled parking to reduce the bin width.

5.4.7 Choice of layout

There are many layout options to choose from[29] but only a small number tend to be used extensively in practice.

Before configuring the internal layout, it necessary to obtain the following information:

- Purpose and intended use of car park
- Static capacity based on a demand study
- Bay turnover and peak hourly flows through the entry and exit barriers
- Position of entry/exits in relation to site and external highway
- Capacity of highway network and its junctions to support peak exit flows
- Topographical survey showing size and shape of site and the existing levels
- Proximity and nature of adjacent streets and buildings
- Pedestrian links to nearby facilities
- Pedestrian needs and accessibility
- History of previous uses of site and any constraints
- Ground investigation showing nature of ground and ground water levels, along with any contamination issues
- Environmental conditions and restrictions
- Drainage and attenuation requirements along with flood risk assessment
- Planning conditions imposed on the development of site
- Preferred structural system and column grid
- Fire considerations and access for firefighting equipment
- Futureproofing (future needs and change of use)
- Any other special conditions e.g., measures to prevent terrorism, restrictions as a result of flightpaths and potential interruptions to airborne communications systems

With this information to hand, the factors that will have the greatest influence on the size, shape and layout can be identified and listed in order of importance. Needless to say, some items will appear to have equal importance and therefore compromises will have to be made in determining which will have priority, but in doing so it is important to remain focused on those that will maximise the functionality of the car park.

Tidal flow will usually occur in car parks that serve business premises or social events, as users park for longer periods and generally arrive at the same time and depart at the same time. However, in some facilities e.g., a theme park, vehicle arrivals can be staggered but departures will be highly concentrated. In car parks for hospitals (including staff parking), train stations etc. a higher turnover of parking bays can occur even though some of the users may park for long periods.

It is important to be able to arrive and depart with minimum delay at facilities that require a fixed arrival/departure time. A single search path of 500 bays is the ideal maximum, and for a greater number there should be more than one search path or additional ramps/shortcuts to move more quickly vertically without the need to circulate each level. Ideally, the layout should be configured to achieve the required static capacity with maximum dynamic efficiency, but it should be borne in mind that when operated close to capacity the risk of internal congestion and delays will increase.

Internal variable message signage or bay guidance systems can be used to manage circulation and to divert drivers past full floors or areas. These systems need to be designed with good visibility and located before the point at which a decision has to be made.

Where the flows are considered to be tidal, the static capacity should be no greater than the dynamic capacity in order to fill or empty the car park within one hour. However, the static capacity can be greater than the dynamic capacity where bays are utilised for long-term parking and not included in the anticipated peak hourly flows.

Sloping parking decks are not suited to where there is high demand for accessible parking and/or users with pushchairs or shopping trolleys. Also, accessible parking bays and bays dedicated to persons with special needs are best positioned nearest to the lifts and stairs and in the quieter areas of the car park, thus avoiding the main circulation routes where traffic movements are intense.

The positioning of main and secondary stair cores is significant, as when situated within the main circulation or adjacent to the vehicle ramps, they can reduce the dynamic efficiency and increase the risk of injury to pedestrians.

The layouts most often seen in practice are illustrated in the following subsections, with a brief note on the advantages and disadvantages of each.

5.4.7.1 Flat deck with internal transverse ramp (Figure 5.13)

One-way flow, clockwise or anticlockwise, which requires a minimum of three bins to accommodate the full-height two-way vehicle ramp. Horizontal external perimeter. All decks are flat with good pedestrian access. Main search path and recirculation is good but there are unavoidable cul-de-sacs. Circulation efficiency is low so unsuitable for high turnover of parking bays. Suitable for static capacity of approx. 500 bays owing to long search path. Gradient on full-height vehicle ramp exceeds recommendations in Table 5.5.

Figure 5.13: Flat deck with internal transverse ramp

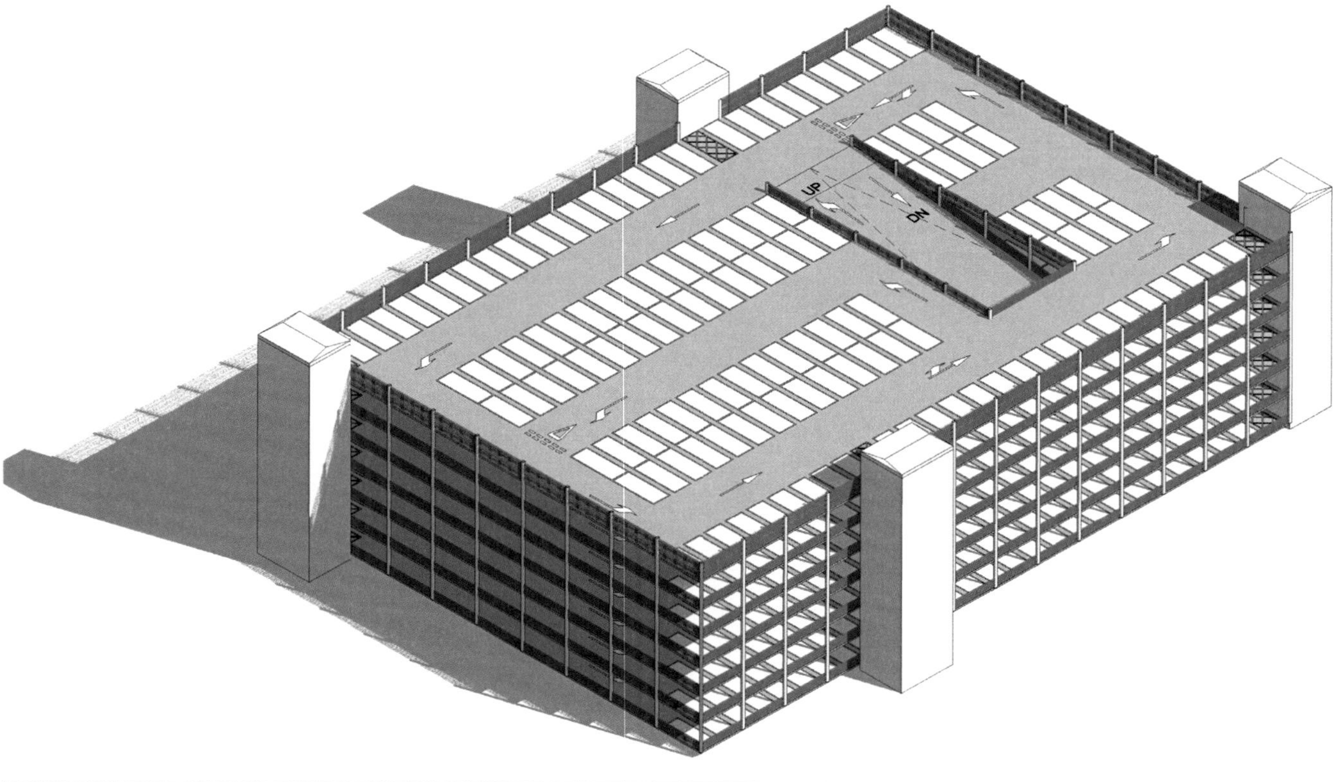

5.4.7.2 Flat deck with internal longitudinal ramp (Figure 5.14)

Two-way flow with a horizontal perimeter on three sides and the full-height parking ramp on the fourth side to form a semi-helix. Both entry and exit routes require the decks to be circulated completely with no shortcuts. Recirculation is poor. Pedestrian access is restricted by the full-length, tapered split-level. More suitable for tidal flow than high turnover of parking bays as the circulation efficiency is poor as a result of vehicles having to cross traffic flow to access opposite parking bays and the 90° turns at the end of the drive aisles. Suitable for static capacity up to 500 bays.

Figure 5.14: Flat deck with internal longitudinal ramp

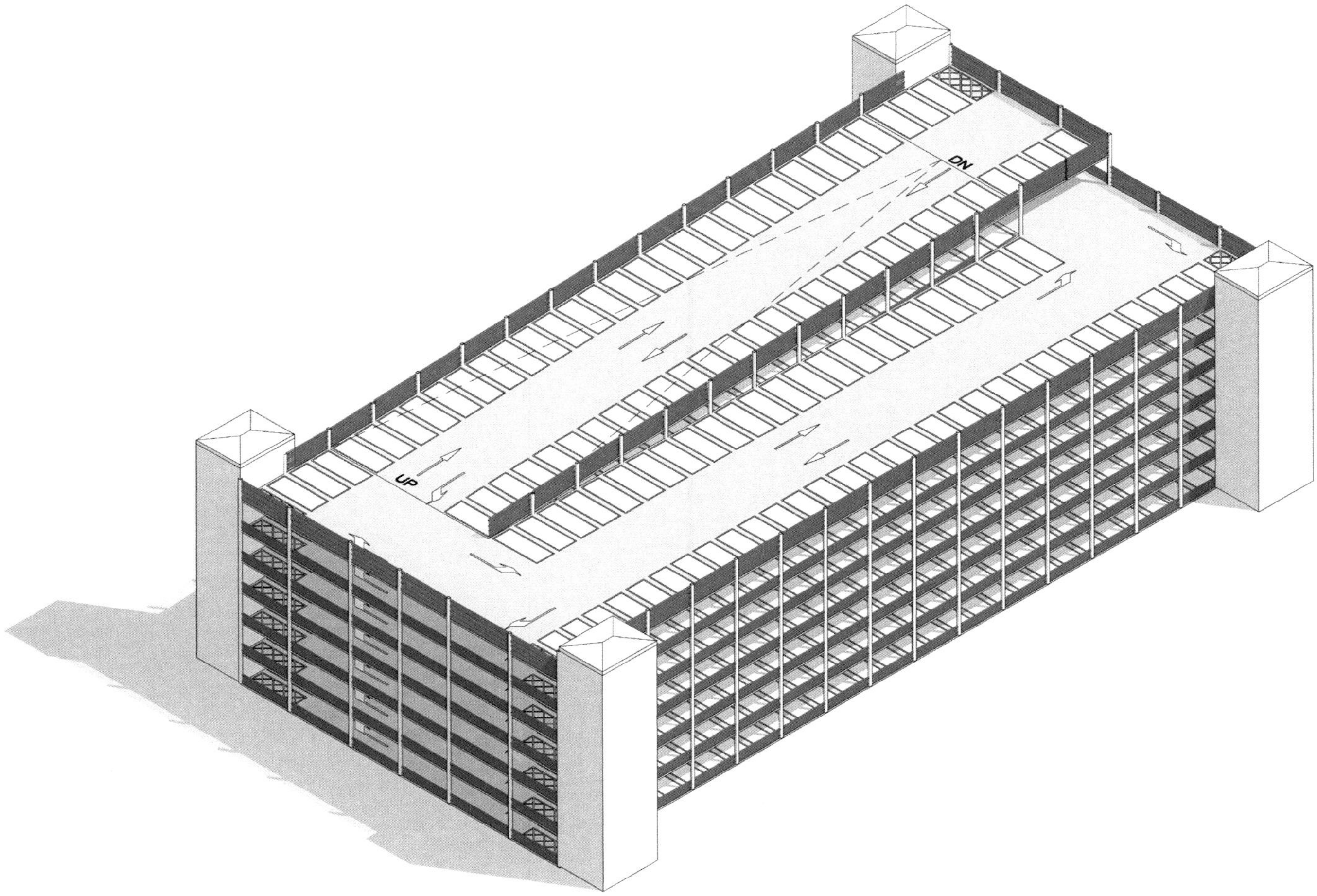

5.4.7.3 Flat deck with integrated D-shaped ramps (Figure 5.15)

One-way clockwise or anticlockwise flow. Horizontal external perimeter. All decks are flat with good pedestrian access. Main search path is good with ability to shortcut main circulation on entry and exit. Circulation efficiency is good, so suitable for large car parks with a high turnover of parking bays. Dynamic capacity is approx. 1,300 vph.

If the D-shaped ramps are made circular and positioned outside of the car park (i.e., clearway ramps which exclude the circulation aisles), the dynamic capacity can be increased to more than 1,500 vph.

Figure 5.15: Flat deck with integrated D-shaped ramps

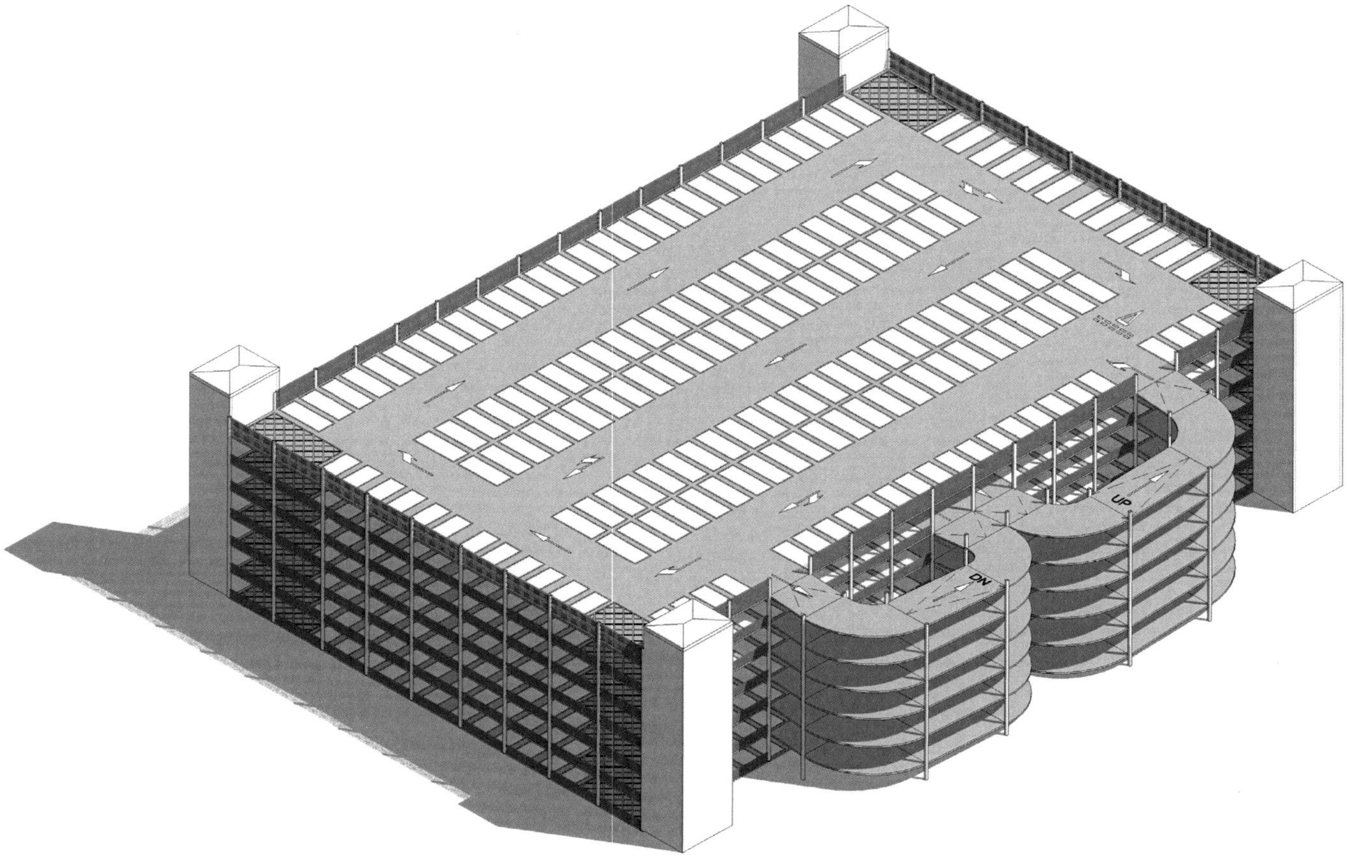

5.4.7.4 Flat deck with two-way circular ramp (Figure 5.16)

One-way flow with flat decks and good pedestrian access. Horizontal external perimeter. Circulation and recirculation is poor as vehicles have to cross opposing flows to access the two-way circular ramp. Unsuitable for high turnover of parking bays. Suitable for a static capacity of approx. 300 bays.

Figure 5.16: Flat deck with two-way circular ramp

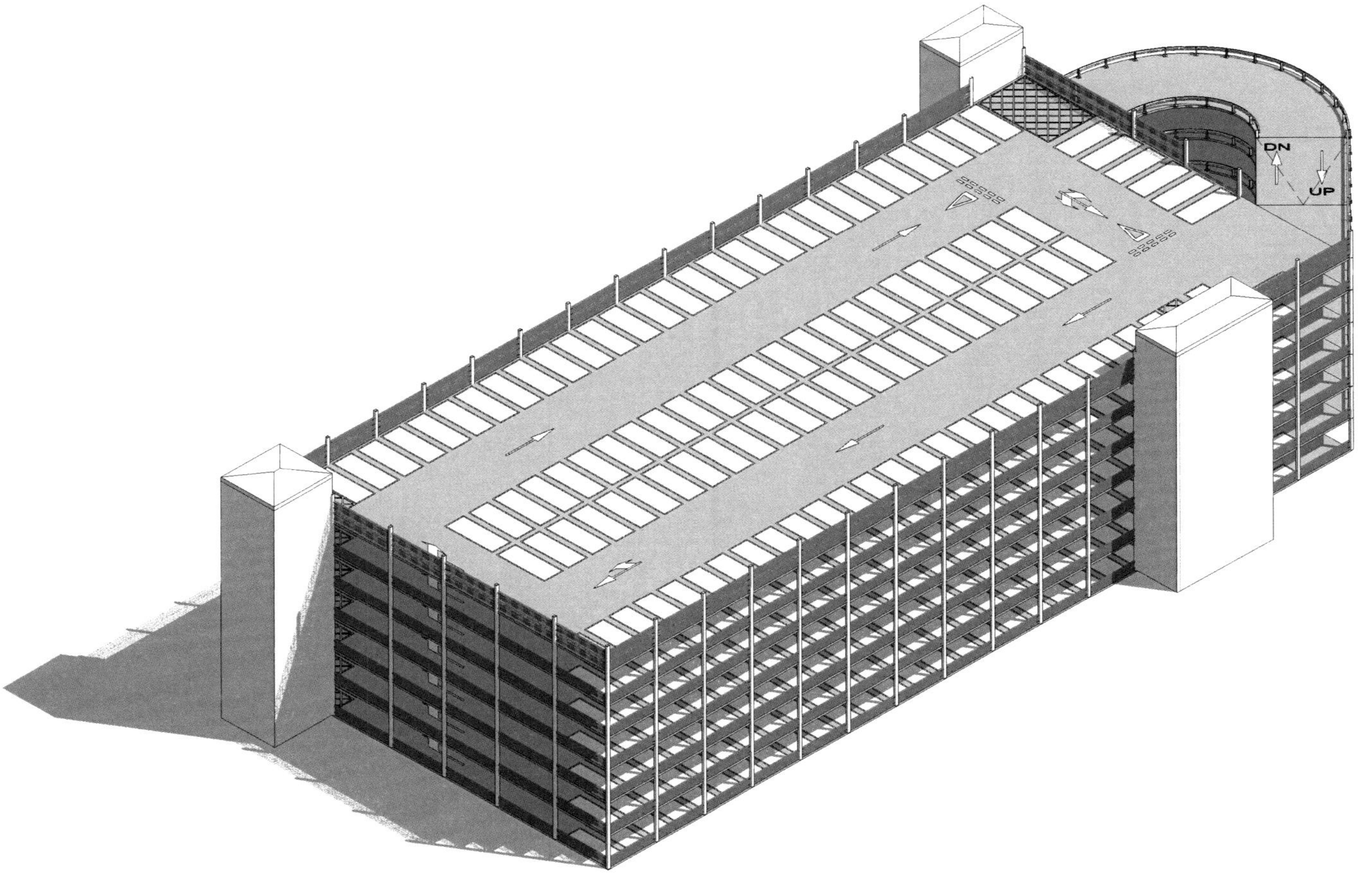

5.4.7.5 Semi-helix with integrated D-shaped ramp (Figure 5.17)

One-way flow with a horizontal perimeter on three sides and the full-height parking ramp on the fourth side to form a semi-helix with an external D-shaped exit ramp. Entry route has a good search path with scope to recirculate via the circular ramp. Exit route is rapid via the circular ramp. Full-length, tapered split-level prevents making a shortcut and restricts pedestrian access between the decks. Suitable for both tidal flow and high turnover of parking bays with a static capacity of approx. 600 bays.

Figure 5.17: Semi-helix with integrated D-shaped ramp

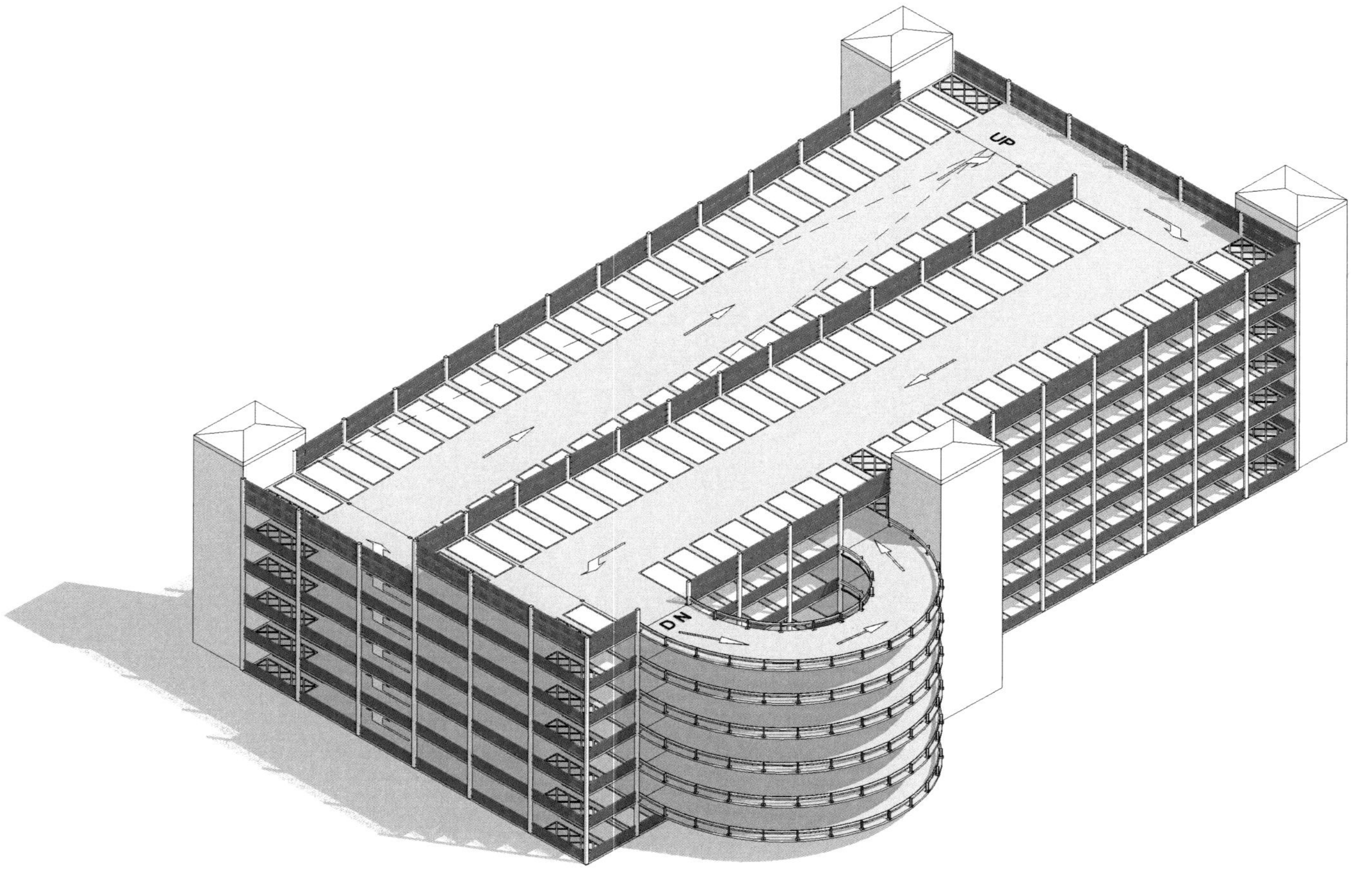

5.4.7.6 Full helix with internal ramps (Figure 5.18)

Two-way flow with horizontal gables and full-height parking ramps on the long sides to form a full helix. Both entry and exit routes require the decks to be circulated completely with no shortcuts. Recirculation is poor. Pedestrian access is restricted by the full-length, tapered split-level. Suitable for tidal flow and low turnover of parking bays as the circulation efficiency is poor due to 90° turns at the end of the drive aisles and vehicles having to cross traffic flow to access opposite parking bays. Suitable for a static capacity of up to 400 bays.

Figure 5.18: Full helix with internal ramps

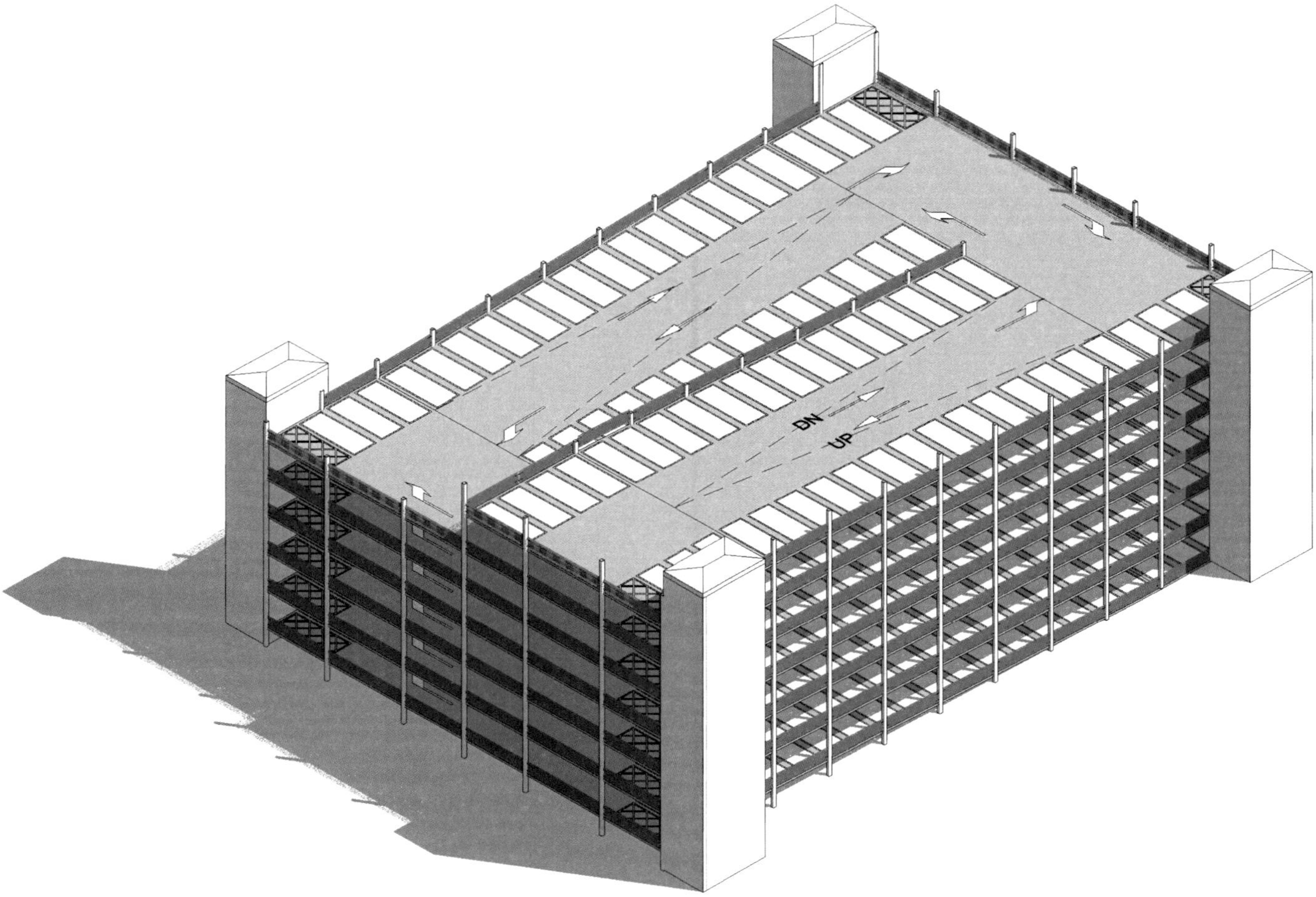

5.4.7.7 Vertical circulation module (VCM) with two-way flow (Figure 5.19)

The VCM essentially comprises two opposing quarter-height parking ramps with a half-height vehicle ramp which connects the parking ramps. Both entry and exit routes require the decks to be circulated completely with no shortcuts. Only one gable end is horizontal. Recirculation is poor. Pedestrian access is restricted by the full-length tapered split-level. Suitable for tidal flow and low turnover of bays as the circulation efficiency is poor, with vehicles having to cross the traffic flow to access bays on the opposite side of the drive aisle, and there are 90° turns at the end of the drive aisles. Suitable for very small car parks with a static capacity of approx. 300 bays.

Figure 5.19: Vertical circulation module (VCM) with two-way flow

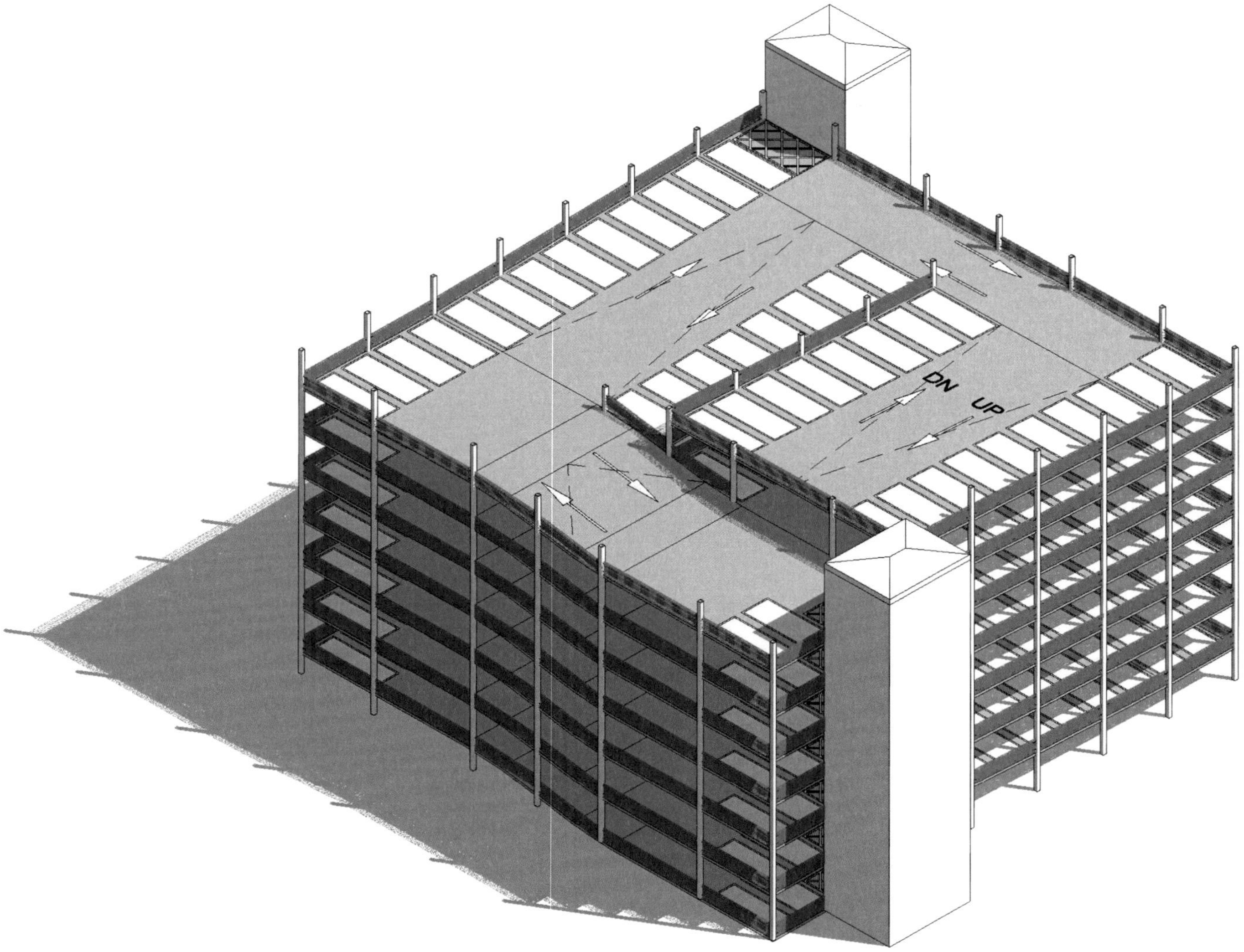

5.4.7.8 Vertical circulation module (VCM) with one-way flow (Figure 5.20)

If two VCMs are butted together at the position of the vehicle ramp, the flow can be made to be one-way clockwise or anticlockwise. Entry and exit routes are separated which enhances circulation and recirculation. It is also possible to bypass the main circulation route to rise quickly through the car park and descend quickly to the exit. Both gables are horizontal, but the sides are sloped along the length of the parking ramp. It is possible to extend the car park horizontally from either gable with level decks which provides good pedestrian access. Suitable for medium and large car parks as circulation efficiency is good. If the two-way vehicle ramp is a minimum of four parking bays wide this layout is suitable for a dynamic capacity of around 900 vph. If a shortcut cross aisle is incorporated into the entry circulation, the dynamic capacity can be increased to around 1,100 vph.

Figure 5.20: Vertical circulation module (VCM) with one-way flow

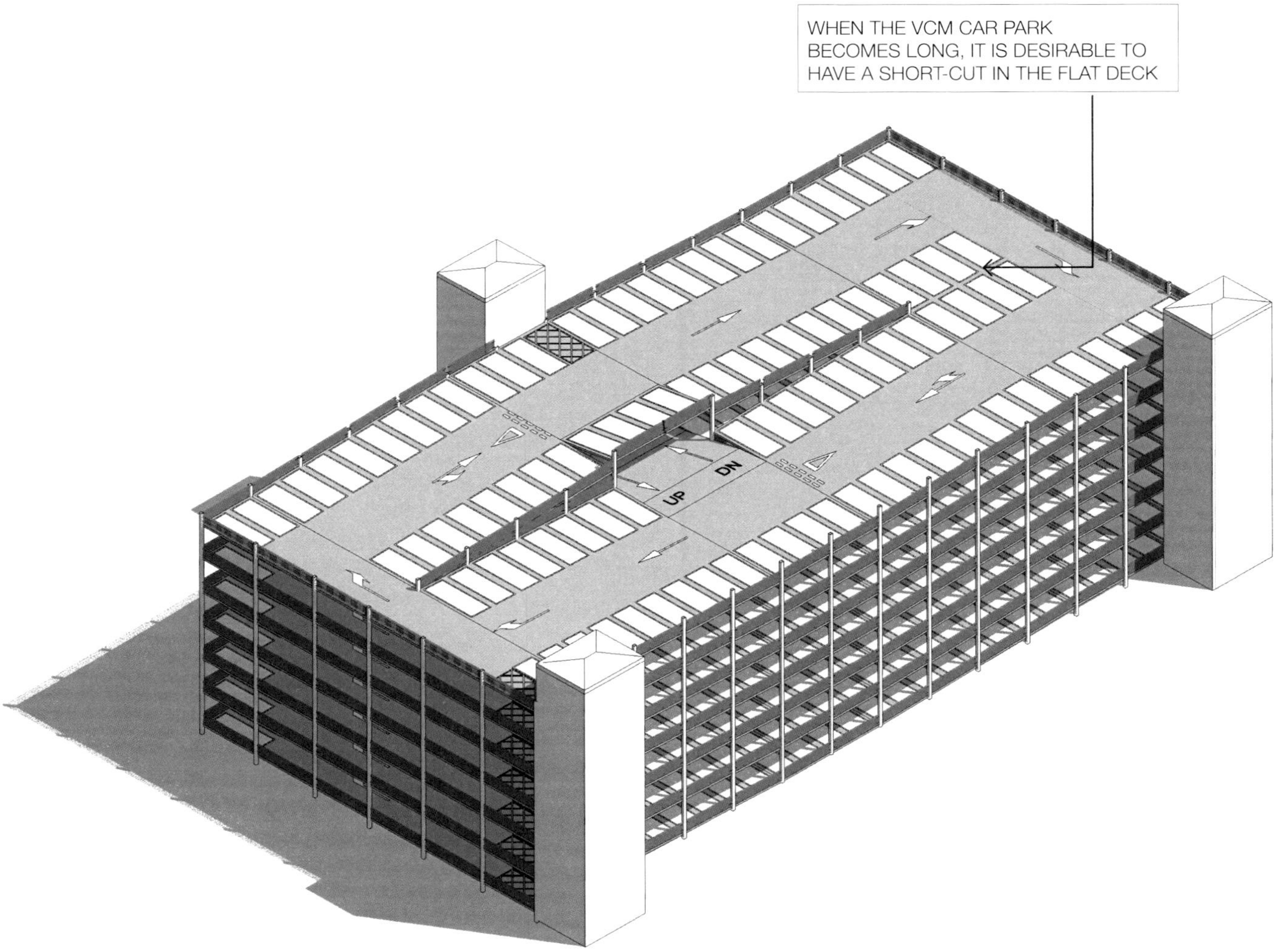

5.4.7.9 Split-level with separated flows (Figure 5.21)

One-way flow, clockwise or anticlockwise, with two single and one double vehicle ramp to circulate the split-level. Horizontal long sides but sloping gables. All decks are flat, but pedestrian access is restricted by the half-height split-level. Both main search path and recirculation are good, and the exit route can be configured to be rapid. Circulation efficiency is average and suitable for tidal flow and medium turnover of bays. Suitable for a static capacity of approx. 500 bays.

Figure 5.21: Split-level with separated flows

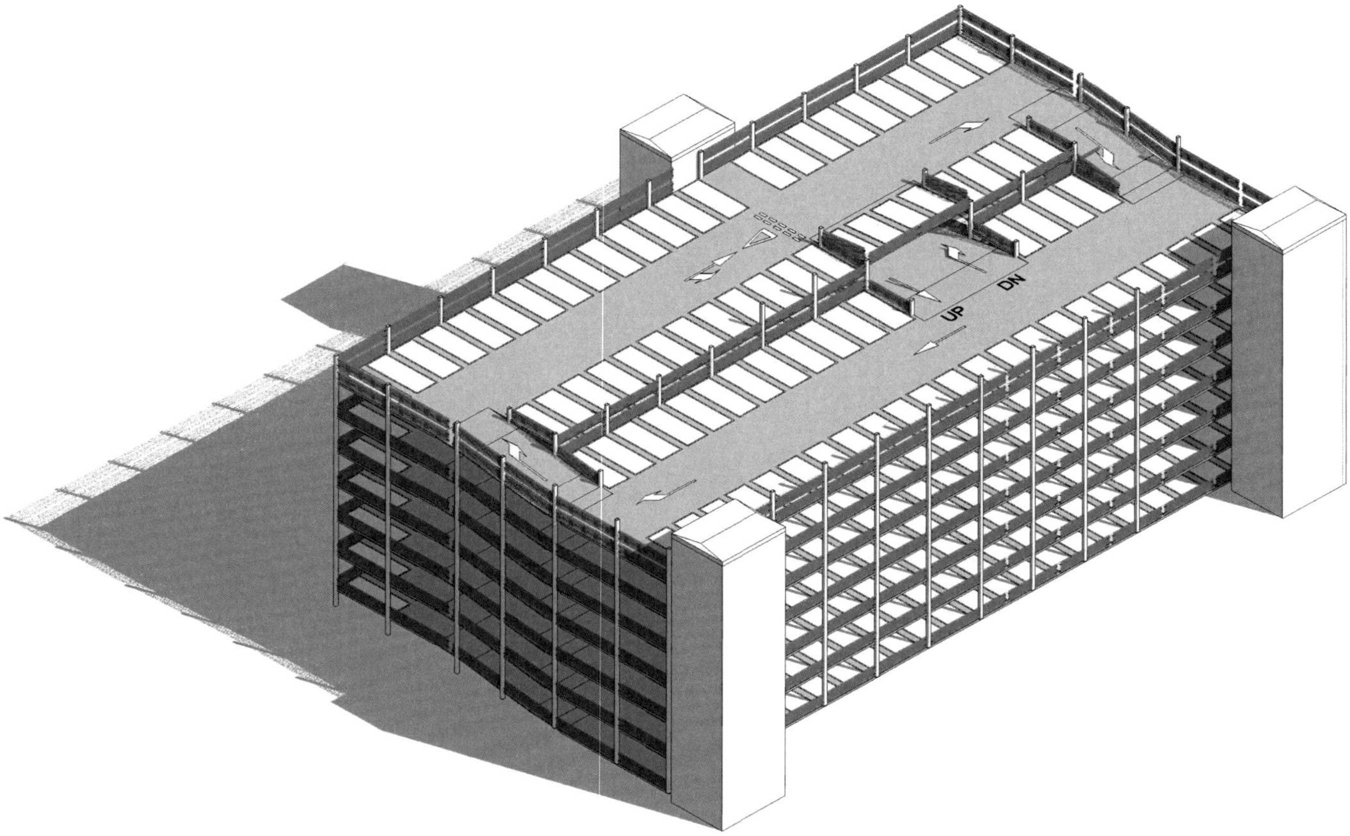

5.4.7.10 Split-level with integrated flows and scissor ramp (Figure 5.22)

One-way flow, clockwise or anticlockwise, with two single ramps and a scissor ramp to circulate the split-level. Horizontal long sides but sloping gables. All decks are flat but pedestrian access is restricted by the split-level. Main search path is long. Circulation efficiency very poor owing to the exit route being contained within the entry route. Recirculation is average and therefore suitable only for very small car parks with tidal flow and low turnover of bays. Suitable for a static capacity of approx. 400 bays.

Figure 5.22: Split-level with integrated flows and scissor ramp

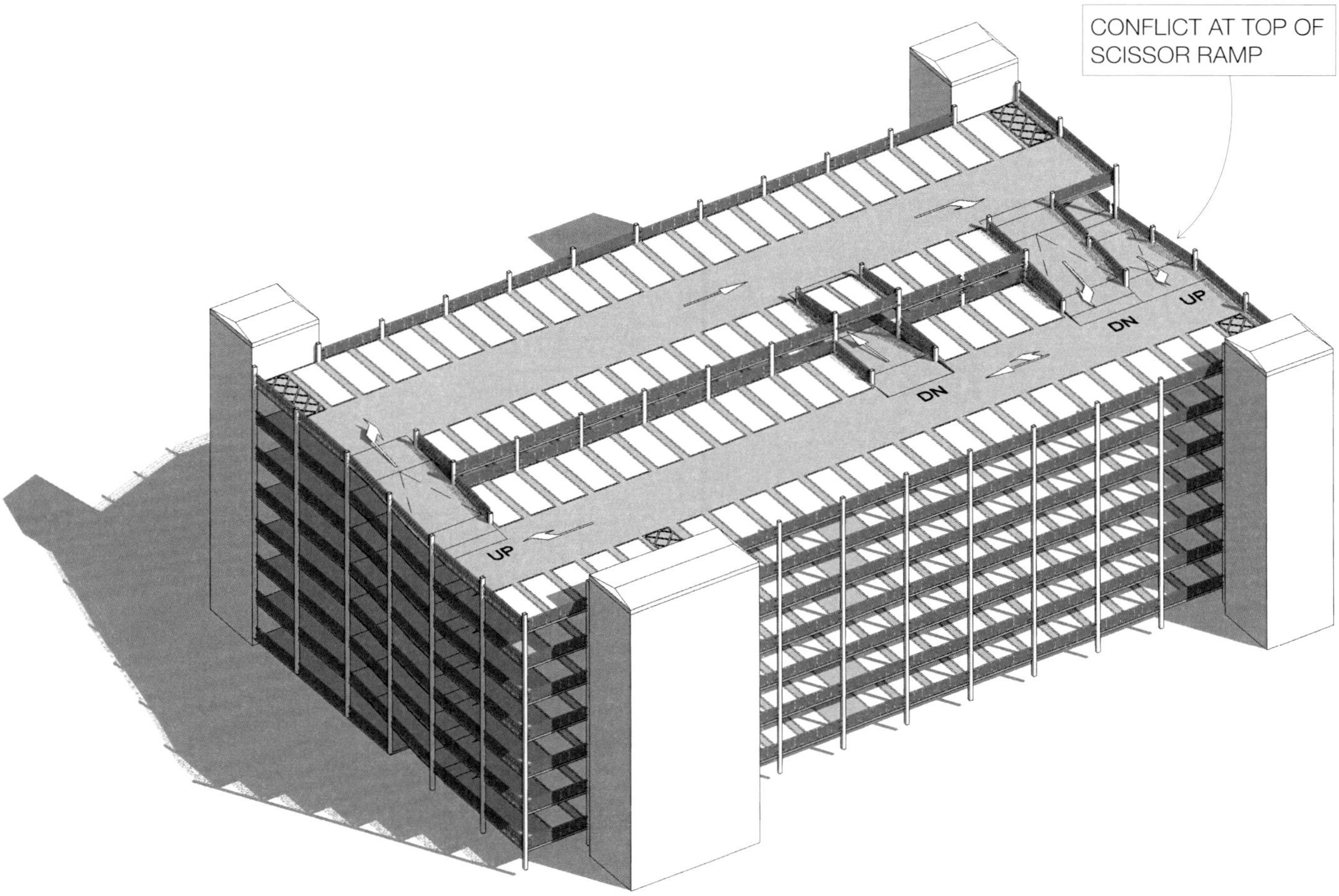

5.4.7.11 Split-level with double scissor ramps (Figure 5.23)

One-way flow, clockwise or anticlockwise, with scissor ramps at each gable end to circulate the split-level. Horizontal long sides but sloping gables. All decks are flat but pedestrian access is restricted by the split-level. Main search path is good, but exiting the car park is slow as decks have to be circulated completely. Recirculation is average and the circulation efficiency very poor owing to the scissor ramps. Layout suitable only for very small car parks with tidal flow and low turnover of bays. Suitable for a static capacity of approx. 300 bays.

Figure 5.23: Split-level with double scissor ramps

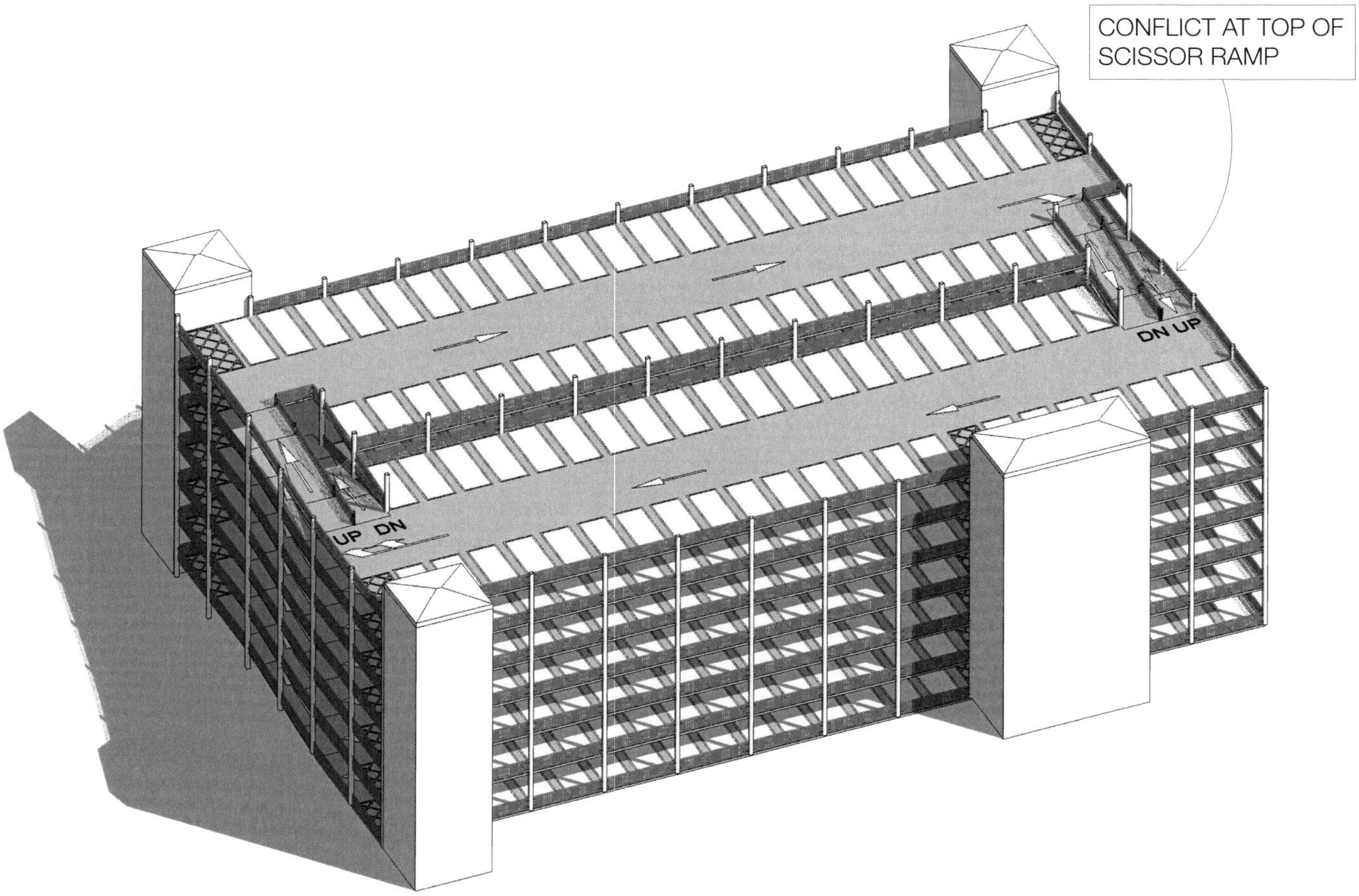

5.4.7.12 Split-level with two-way ramps (Figure 5.24)

Two-way flow with a double vehicle ramp at each gable end to circulate the split-level. Horizontal long sides but sloping gables. All decks are flat but pedestrian access is restricted by the split-level. Main search path and recirculation is average, and the exit route is slow as vehicles have to circulate all decks completely. Circulation efficiency is poor as vehicles have to cross traffic flow to access opposite parking bays. Suitable for tidal flow and low turnover of bays. Suitable for small car parks with a static capacity of up to 400 bays.

Figure 5.24: Split-level with two-way ramps

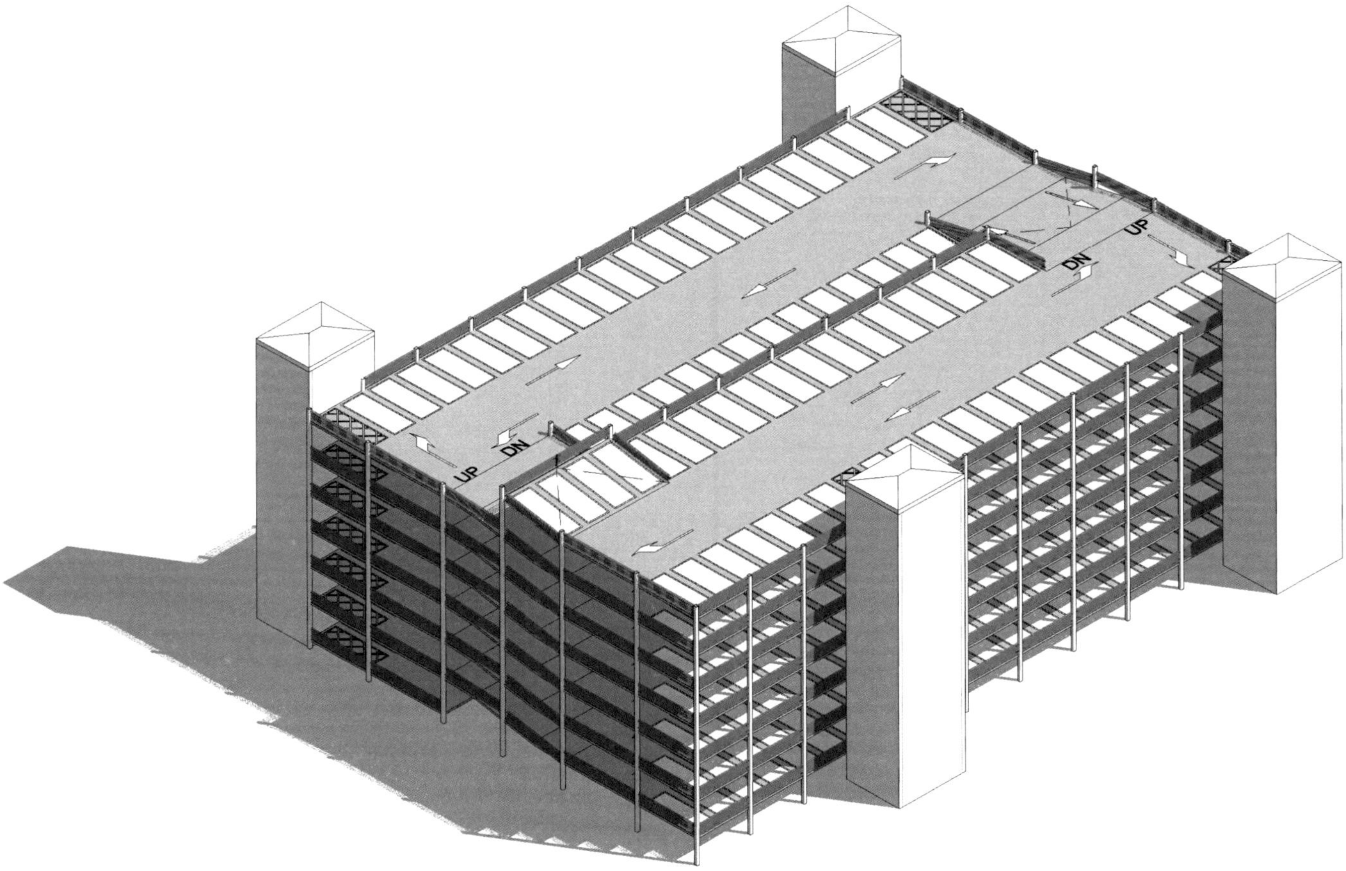

6 Integration of building services (conceptual design)

6.1 Introduction

The design of building services is usually the remit of the mechanical and electrical (M&E) design consultant.

The purpose of this chapter is to highlight key aspects of the building services that require consideration during the conceptual design stage. It focuses specifically on those aspects/activities which are likely to interface directly with the structural and architectural design and/or have a significant impact on overall design solutions.

6.2 Design activities

The key activities for the conceptual design stage should include:

- Reconciliation of client brief against building design
- Building load assessments
- Appraisal of incoming utility requirements
- Assessment of plant area requirements
- Development of services strategy to permit planning of plant rooms and risers
- Development of services zones (e.g., lighting, sprinklers, ductwork) required in each area, to permit planning of structural heights
- Input into facade thermal performance benchmarking (in relation to energy conservation)
- Formalisation of the project's energy strategy (in conjunction with the sustainability strategy)
- Consideration of outline plant replacement strategies
- Building Regulations Part L[33] preliminary assessment (if applicable)

6.3 Engineering services strategy

It is important that the design criteria to be applied to the engineering services is set out at an early stage. This provides a basis for calculating load assessments and system duties to allow the overall services strategies to develop.

The following six subsections consider the key systems. A short note with regard to their interface with structural/ architectural design is supplied for each.

6.3.1 Mechanical services

6.3.1.1 Heating

- Heating is likely to be considered in the context of frost protection for the primary car park space (since it is typically considered unoccupied), but ancillary spaces may require consideration of environmental conditions if occupied

6.3.1.2 Ventilation

- Every effort should be made to take advantage of opportunities for natural ventilation. At each level, openings should have an aggregate area of at least 2.5% of the area of the parking space at that level and be so distributed as to provide effective cross-ventilation
- For all underground car parks (and for those above ground where adequate natural ventilation cannot be provided), mechanical ventilation will be required. Ventilation and smoke ventilation systems have significant implications in terms of spatial requirements, cost, and considerations for secondary electrical power sources

- Smoke ventilation systems may be required to provide enhanced air change rates to cater for the risks associated with electric vehicle fires. These ventilation requirements should be considered at an early stage, in conjunction with the overall fire strategy

6.3.2 Electrical services

6.3.2.1 Lighting

- Appropriate lighting is an extremely important aspect of the design for car parks. This will be required to ensure safe movement of vehicles and pedestrians and to assist in creating an environment that feels safe for users and staff, and helps reduce vandalism and other crime
- The nature of car park structures usually inhibits the scope of the lighting designer, but nevertheless an early association with the architect and structural engineer may produce, within a reasonable budget, a better-considered, visually attractive, easily maintained lighting scheme. An early engagement to influence low headroom and structural form will assist in tackling the common challenge of uniformity, and may provide opportunities for improved lighting efficacies
- External lighting installations should also be considered at an early stage, both from a functional and architectural perspective, as the integration of external lighting is likely to require coordination and interface with structural build-ups, facade detailing and landscape proposals

6.3.2.2 Power

- Historically, car park facilities have not been power-intensive spaces. With the growth of electric vehicles and their associated charging requirements however, power demands are increasing dramatically. This will need early consideration in conjunction with the review of overall maximum demand load assessments to ensure that the appropriate distribution equipment and plant rooms/substations are allowed for
- With overall maximum demands assessed, it is important to engage with the District Network Operator (DNO) to ensure adequate capacity can be accommodated on the local network. Any network upgrade requirements are best identified at an early stage, as they are likely to have significant cost implications and will need to be factored into the cost planning
- Power containment system routing should be considered to minimise exposed cabling, protect against vandalism and coordinate requirements with structural penetrations through beams or slabs
- In conjunction with developing the energy strategy, consideration should be made for the inclusion of photovoltaic (PV) systems and battery storage opportunities — and the benefit these might afford in managing the power demands of the development

6.3.2.3 EV charging facilities

- Both current and future demand for EV charging should be taken into account; this is likely to include the installation of EV chargers with the development of new car parks, along with electrical infrastructure to enable the installation of chargers to an increased proportion of car parking spaces in the future

6.3.2.4 Fire alarm systems

- It is likely that an automatic fire alarm system will be required to meet the requirements of the fire strategy. In most enclosed car parks, this will be recommended to provide early warning to persons within the facility
- The type and arrangement of detection and alarm should consider the expected environment (e.g., exhaust fumes, resistance to vandalism) and be designed to minimise false activations
- Interfaces with other systems (e.g., smoke ventilation, lifts, battery storage, EV charging) will also need to be considered to manage appropriate cause and effects scenarios in the event of fire

6.3.2.5 Lightning protection/earthing

- A risk assessment should be carried out to determine the requirements for lightning protection. These systems can take various forms and invariably will include a network of copper tapes linking air termination to earth termination points. Early engagement with structural engineers can identify opportunities to utilise building structure (e.g., steel columns, steel reinforcement in concrete) and minimise the visual impact of these systems
- In additional to general earthing and bonding requirements, consideration should be given to the need to provide earthing for secondary electric sources (e.g., generators, battery storage, photovoltaics), and the provision of surge protection devices to protect the power and IT equipment

6.3.3 Public health services

6.3.3.1 Drainage

- A well designed and properly maintained drainage system is essential to ensure that all water deposited on the exposed surfaces is rapidly discharged. Early engagement with the structural and architectural designs will be required to ensure falls, drainage channels etc. are designed to suit the type of structure and take account of parking areas, ramps and circulation areas, and other functional areas of the building
- Integration of piped drainage systems should consider both protection from vehicle impact and access for rodding, as well as potential for vandalism. Early engagement with the architect to locate toilet facilities with a view to achieving vertical continuity where possible, will assist to simplify drainage systems and minimise horizontal distribution of piped services
- Also refer to Section 8.6.1

6.3.3.2 Water

- Early design considerations should focus on setting the brief requirements for any water services, particularly where this may be required to integrate with landscape design (e.g., irrigation requirements, water features) or functional areas of the development such as toilet facilities, welfare/maintenance provisions etc.

6.3.4 Fire protection services

Specific requirements for fire protection services should be developed in conjunction with an overall fire strategy, but key aspects are identified in the following two subsections. All have significant space and design implications that require early consideration in the design process.

6.3.4.1 Sprinklers

- Key elements relate to the hazard category of the system, water storage requirements (tank size, water infill arrangements), sprinkler pump types (electric/diesel), primary/secondary power supply requirements, and sprinkler head densities
- Additional consideration should be given to the type of system; in relation to both the expected fire load (e.g., presence of oil, electric batteries etc.), and the environment in which the system would be installed (e.g., external risk of freezing, resistance to vandalism etc.)

6.3.4.2 Fire suppression

- For certain areas of the building, traditional wet sprinklers may not be appropriate, and alternative means of providing fire suppression may be required. For example, electrical switchrooms would generally require an alternative solution to remove the risk of water from electrical equipment. In such cases, a gaseous suppression system may be more appropriate, and this should be identified as early as possible in the design, in conjunction with any alternative fire engineered strategies

6.3.5 Vertical transportation

- Requirements for lifts and escalators will vary significantly between different developments, but multi-storey car parks will usually be required to satisfy legislation for inclusive access and to meet the reasonable requirements of users, e.g., those carrying shopping, children etc. Consideration should be given to the function of the space and whether the lifts are adjacent to manned points
- If a firefighting lift is required, a secondary power supply will be required. This can be significant and will usually take the form of a standby generator, although opportunities may exist to explore a secondary source from an alternative low voltage supply if it can be demonstrated to be independent from the primary power source

6.3.6 IT, security and specialist services

- Requirements for these systems will vary significantly between developments, but it is important to consider the scope that may be required in order to make provision for cable containment systems, power and data requirements. Interfaces may be required to structural or architectural elements to accommodate equipment such as CCTV, or access control (vehicle or pedestrian), and depending on the extent of system requirements there may be a need for central control equipment, or communications rooms which will need to be accounted for within the architectural planning

6.4 Sustainability and energy position

In the UK, depending on the size and complexity of a project, the construction programme often requires a development to be futureproofed against changes to Building Regulations that are likely to occur during the period between design and construction. A review of the current benchmark targets should be undertaken to set energy targets to achieve appropriate emissions reductions against Building Regulations Part L[33], combined with improved fabric performances. These fabric performances are key to reducing the energy required by building services systems (primarily heating and cooling) and therefore active engagement by the building services engineer at the earliest opportunity is desirable.

Car parks are generally classed as unoccupied buildings, and with no heating and cooling they do not fall into the scope of the Building Regulations Part L. A Simplified Building Energy Model (SBem) calculation is not required. However, designers should ensure that in these areas, lighting efficacies (LM/W) and specific fan power (w/(ltr/sec)) are within the requirements of Part L, and that every effort is made to reduce the energy requirements of the building.

As part of a planning requirement, any development will require the submission of an energy strategy. The Greater London Authority's *Energy Planning Guidance*[34] (and particularly its *Energy Assessment Guidance*) is a useful resource — while Chapter 8 of this guidance also provides more detail.

6.5 Plant replacement

Consideration of services distribution zones is key to determining the structural heights within the car park. The building services engineer must provide early input into the zones required to accommodate, for example, lighting, sprinklers, ductwork and services distribution, to allow these to be considered in setting the overall slab-to-slab heights to achieve the clear height requirements for vehicles. Typically, it will be these services elements that provide the high level constraint on height in a completed car park. Section 5.3.8 of this guidance discusses headroom in further detail.

In relation to clear heights and as a core aspect of creating the engineering services strategy, outline solutions for the movement of primary plant need to be developed. It is reasonable to assume that plant can be broken down into component parts for both installation and removal and it is expected that specialist attendance may be required to carry out assembly, disassembly, and plant movement as necessary. However, it is preferable for initial plant placement to deliver entire units wherever possible.

The structure and architecture need to be designed to accommodate the weights and clearances of building services elements, ideally without the need to remove any structures or fixed elements of the building fabric.

Building services equipment will typically need to be renewed every 15–20 years, i.e., two or three times during the life of a car park. It should therefore be designed and installed so that it can be removed and replaced without damaging or altering the structure. As such, a surface installation fixed directly to the structure should be considered.

Consideration should also be given to the operational requirements of the car park – e.g., if it will not be possible to close the car park for an extended period to replace building services plant, the design should allow the car park to be closed section by section while the rest remains operational.

Key elements in determining the plant replacement proposals at conceptual design stage are to set the primary plant positions, consider strategies for horizontal and vertical movement of primary plant, and to set out clear height and width requirements for primary access routes and doors. Any permanent lifting beams (e.g., at the head of lift shafts) should be considered so that these can be integrated with the structural design, although typically, where individual items of plant require lifting equipment, these are often accommodated by use of temporary lifting beams erected and certified by the contractor/specialist carrying out the plant replacement. Additionally, any third-party plant and personnel access requirements should also be considered. For example, UK Power Networks (UKPN) typically require 24hr unrestricted access to their substation locations, with routes and clearances used as the basis for lease agreements.

7 Fire considerations (conceptual design)

7.1 Background

Designing for fire should be integrated into the design process from the start, as it fundamentally affects the layout, materials choice and operation of the car park.

Fire considerations are broad and include:

- Controlling fire risks
- Means of escape
- Structural response
- Fire spread
- Facilities for fire brigade
- Property protection
- Insurance

Current guidance is based on tests using outdated vehicle types. Vehicles have become substantially larger, which equates to a larger fire load. Greater use of plastics and synthetic materials may mean higher smoke density and toxicity. Plastic fuel tanks present risk of rupture and the potential for a running fuel fire. Vehicles with alternative fuel sources such as LPG present another risk, and there are risks arising from rapid vehicle charging and from lithium ion batteries in electric and hybrid vehicles, most notably thermal runaway of the batteries. Designs should consider these issues.

Some recent fires have demonstrated that current design standards may not be sufficient to produce a fire-safe car park, in certain circumstances. Therefore, it is critical that designers do not unquestioningly follow design codes without first considering whether the approach is suitable under the circumstances of the particular project.

A CROSS Safety Alert[35] was issued in February 2018, highlighting issues arising from the Kings Dock fire which occurred on New Year's Eve 2017. The severe fire at the 7-storey Liverpool Echo Arena multi storey car park gutted the building, and up to 1,400 cars were destroyed. The fire was unusual in that it rapidly spread from a single car, but the outcome was disproportionate to the cause. The Alert draws attention to the fact that such events could happen again, and lessons should be learnt when assessing the risks in new car parks (Figure 7.1).

Figure 7.1: Liverpool Echo Arena MSCP fire, Liverpool, UK

7.2 Legislation

Consideration must be given to the requirements of national and local authorities having jurisdiction. For example, the enforcing authorities throughout the UK are the local authorities and the fire authority. Their powers rest in the Building Regulations[10, 36, 37], and the Regulatory Reform (Fire Safety) Order 2005[38] respectively. Local authorities may have additional powers under local legislation.

'Statutory guidance' is often written to assist designers in complying with regulations, and reference made to other standards and guidance.

Standards and statutory guidance are usually minimum standards to be complied with. Designers should look to go beyond the minimum where practicable — and the information contained in this guidance is considered to be good practice.

7.3 General principles

Designing for fire should start with a fire strategy that sets out the principles of the design. As many aspects are interlinked, the fire strategy will inform all areas of the design. A fire strategy can be written by an architect or other designer, but where the design deviates from accepted codes of practice, a specialist fire engineer may need to be part of the design team. In all cases the designer making choices about the fire aspects of the project will need to be competent to do so.

The risk from fire depends on several factors, including the fire load and heat output — which are influenced by the type and number of vehicles, sources of ignition, layout, size and shape of the fire compartment, ventilation and the ability to extinguish or suppress the fire. A prime source of fire ignition is arson, and good security design can help to reduce this.

Fire spread from vehicle to vehicle is a major concern when considering fire loading in a car park. Research conducted by the Building Research Establishment on behalf of the UK Government[39] into the growth and spread of fires in car parks revealed that there is a risk of fire spreading from car to car, particularly where early intervention by the fire brigade is not possible. The research showed there was a particular risk in split-level car parks with bays nose to nose across the levels.

The fire loading will depend on:

- Number and mix of vehicles in the car park at the time of fire
- Proximity of one vehicle to another
- Ceiling height
- Degree of ventilation in affected area
- Presence of sprinklers

The general principles of means of escape are that an occupant should be able to turn their back on a fire and move to a place of relative safely within a reasonable time, and then to a place of ultimate safety away from the building. The period immediately prior to this is known as a 'pre-movement time', which depends on the ability of the occupier to become aware of a fire and to react. Good sightlines and an automatic fire alarm system will help to minimise this time. Escape distances and layout of escape routes should enable occupants to have multiple options for escape, and if not, the routes should be very short. Number of occupants, number of escape routes, and width of doors and stairs will impact on the evacuation time. It is essential that account is taken of all occupants including children, the elderly, and those with sensory or mobility impairments. An assessment of the likely occupant characteristics should be undertaken at the outset. Wherever possible, designs should be such that all occupants can self evacuate without assistance.

In order for the building to be adequately managed throughout its life, the fire safety information should be updated to incorporate any changes through construction and given to the operator of the car park. This needs to include key information about the fire safety strategy including guidance on how to maintain fire safety equipment and services.

7.4 Specific issues

The use of vehicle ramps as escape routes should not be part of the design. Ramps are generally too steep to be treated as suitable escape routes, and vertical escape routes should be via protected stairways.

Structural fire response should be considered in the context of the assessment of the fire loads within the car park, together with disproportionate collapse assessments. Codified structural fire resistance ratings are based on standardised ISO fire curves and may not reflect a real fire situation. It is recognised that well ventilated spaces do not heat the structure to the same extant as enclosed spaces. Some of the available literature suggests structural fire resistances as low as 15 minutes for well ventilated car park structures. This should be used with extreme caution, as a fire which spreads between vehicles is likely to produce a sufficient heat release to potentially precipitate a collapse in a building with such a low fire rating.

If the car park is part of another building, the fire risks of the car park to the rest of the building should be considered. This is especially pertinent in residential settings where people may be asleep, and there may be a 'stay put' evacuation strategy. Where a building containing a car park requires sprinklers for the structural fire resistance rating, they should always be installed in the car park. Basement car parks require smoke clearance systems. Where this is a mechanical system, sprinklers should be provided.

Fire service intervention should never be assumed for structural fire resistance, fire spread or means of escape unless part of a probabilistic risk assessment to a recognised code of practice. In particular it should not be assumed that fire services will rescue mobility-impaired persons; rather the building should have suitable measures for self-evacuation, or appropriate management measures in place. Fire service attendance times will vary considerably by country and local situation, e.g., urban versus rural areas. Nonetheless, adequate facilities need to be provided for the fire services. This should include good access to and around the site, taking into account security measures such as gates and barriers. Access within the car park to enable fire services to reach the source of the fire is very important. This might include firefighting stairs/shafts and smoke control systems for ventilation. Water supplies such as hydrants and provision for dry risers and sprinkler tank infills are crucial. Controls should also be provided for ventilation systems and to shut off electrical systems.

Where the car park is close to the site boundary, it is necessary to ensure that fire will not spread to an adjoining building, and will not be at risk itself from a fire on an adjacent site. Open-sided car parks by their very nature do not have fire-resisting enclosures, so the distance to the boundary becomes the critical factor. Care needs to be taken if any combustible materials are used in the facade of the car park, such as timber cladding or green walls, as these can contribute to fire spread.

Car parks often have ancillary uses such as cycle parking, ShopMobility[40] and plant areas. There is a growing trend towards cycle repair areas and parking for alternative transport modes such as electric cycles and scooters. Consideration needs to be given to the risks these different uses pose. Generally, fire compartmentation is required between different uses. Some facilities (e.g., electricity substations) may need a higher standard of fire resistance than the general car park, and the structural fire resistance in that area should reflect that.

8 Sustainability and environmental considerations

8.1 Introduction

This introduction serves as a backdrop to the key sustainability and environmental issues that structural design engineers face. Global warming needs to be limited to 1.5°C above pre-industrial levels to avoid the worst effects of climate change, requiring unprecedented changes globally[41]. The UK Committee on Climate Change shows current targets to reduce carbon emissions from countries globally would lead to warming of 3°C by 2100. The UK was one of first countries to declare a climate emergency, setting a target to achieve net zero carbon emissions by 2050, and is a signatory of The Paris Agreement, pursuing efforts to limit a global temperature rise to just 1.5°C above pre-industrial levels. Delivering 'net zero' emissions will require drastic action across all sectors of the economy.

Current planning policies and regulatory requirements in the UK address aspects of net zero carbon emissions. At a national scale, the *National Planning Policy Framework (NPPF)*[8] includes requirements to reduce operational carbon emissions through building design and maximise the generation of renewable, low carbon and decentralised energy and heat. Approved Document L of the Building Regulations sets standards for the minimum energy performance of new and existing buildings[33]. At the local scale, many planning authorities include energy performance and operational carbon reduction policies in their local plans and associated documents, often requiring an improvement on the Building Regulations Part L minimum performance requirements. In London, the Greater London Authority (GLA)'s *London Plan* includes policies that cover operational and embodied (whole lifecycle) carbon emissions, circular economy strategies, in-use energy reporting and greening of urban areas[34].

Nature is also declining globally at rates unprecedented in human history and the rate of species extinctions is accelerating; average global temperature increases of only 1–2°C (above pre-industrial levels) could commit 15–40% of species to extinction[42]. It is not too late to make a difference, but only if we start now at every level — from local to global — through 'transformative change' to conserve, restore and utilise nature sustainably. The UK Government published the *25 Year Environment Plan*[43] in 2018, which sets out what will be done to improve the environment in a generation. As part of the Environment Act 2021[44], updates to national planning policy include a requirement for 'biodiversity net gain' for developments.

The following sections provide more specific narrative and examples that can be related to car park design.

8.2 Net zero carbon

A net-zero carbon asset (including multi-storey or underground car parks) is one where the sum total of all asset-related greenhouse gas (GHG) emissions, both operational and embodied, over its lifecycle including disposal + offsets = 0. Embodied carbon (EC) and operational carbon (OC) may be defined as follows:

- Embodied carbon (EC): The GHG emissions associated with materials and construction processes throughout the whole lifecycle of an asset (Modules A1–A5 'product sourcing and construction stages', B1–B5 'use, maintenance, repair, replacement and refurbishment stages' and C1–C4 'end-of-life stages')
- Operational carbon (OC): The GHG emissions arising from all energy and water consumed by an asset in use, over its lifecycle (Modules B6 and B7 'operational energy and water use')

Achieving net zero carbon for a multi-storey or underground car park would require calculating and reducing the impact over its lifecycle, before offsetting the residual emissions. The hierarchy of net zero design[45] should be followed to achieve this (Figure 8.1).

Figure 8.1: Hierarchy of net zero design

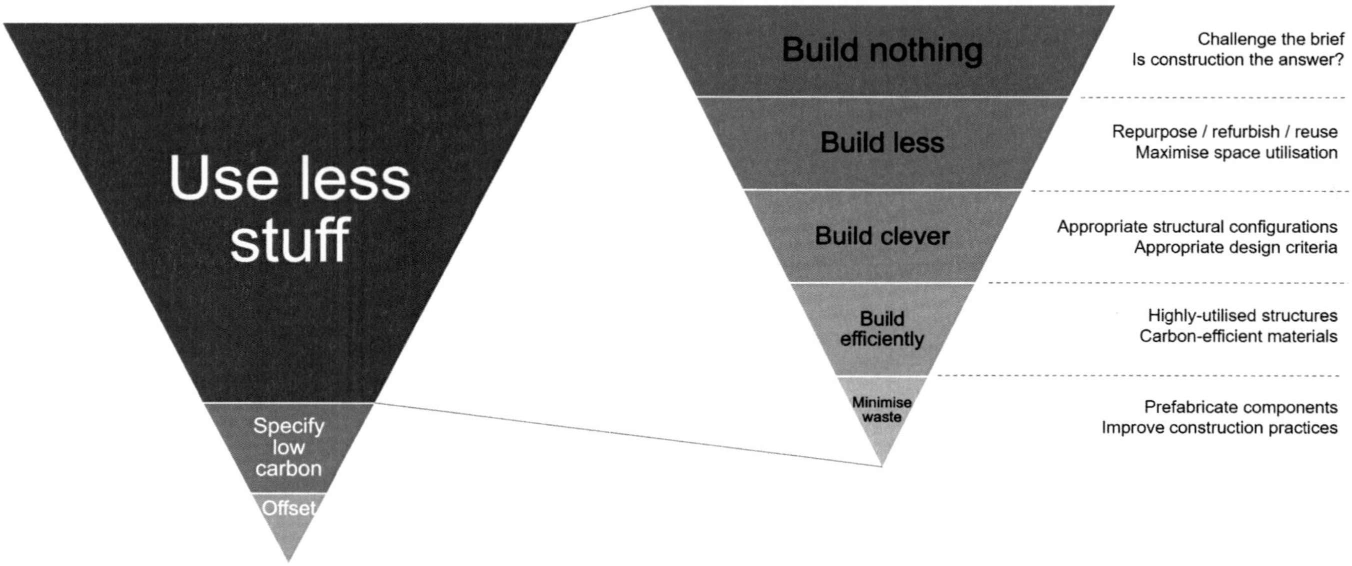

In practice, key considerations for reducing embodied carbon associated with multi-storey and underground car parks should include:

- Retainment, refitting and refurbishment of existing structures where feasible, following prioritising materials reclamation and reuse, before removing from site for remanufacturing or recycling
- Minimising quantities of materials required, by helping the client to develop a brief that is efficient in terms of how much new structure is being asked for, by designing structures with appropriate configurations and direct load paths, and ensuring that those structures are highly utilised when final detailing is undertaken
- Pursuing innovative design solutions and low-carbon materials
- Reducing carbon intensity of final specified materials as far as possible[46–49]
- Offsetting any residual carbon emissions

Embodied carbon in structural materials is typically due to:

- Burning fossil fuels to create heat in steel furnaces and cement kilns
- Chemical reactions in the creation of cement
- Use of electricity (which will be decarbonised in the future)

Constructing car parks uses energy and results in carbon emissions, as well as using a range of other materials and resources. Once a structure has come to the end of its life, these resources are still potentially available for use. Thus, in addition to reducing embodied carbon we must consider the resources as a 'store' rather than a 'flow' and structures should be thought of as 'material resource banks'[12].

This results in two drivers for design: minimising emissions today in line with Fig. 8.1, and minimising emissions in the future by designing our material resource bank to enable reuse in the circular economy (Section 8.3). While often seen as opposing, both can and must be achieved.

A guidance document for calculating embodied carbon for structural components has been published by The Institution of Structural Engineers[50]. This guidance is applicable to infrastructure and buildings, including car parks. Other useful guidance documents include the RICS' professional statement: *Whole life carbon assessment for the built environment*[51] and BS EN 15978[52].

Achieving a whole lifecycle net zero carbon car park would also require operational carbon to be reduced to net zero as follows:

- Reducing energy demand as far as possible
- Meeting energy demand as efficiently as possible

- Generating as much energy on site as possible
- Using off-site renewable energy sources as much as possible
- Offsetting any residual carbon emissions

Conceptual design structural engineering principles which impact operational carbon are also outlined in Section 8.5.

8.3 Circular economy

There is a large amount of overlap between actions that may be taken to reduce embodied carbon, and what is known as the 'circular economy'[12]. A circular economy can be defined as one where materials are retained in use at their highest value for as long as possible and are then reused or recycled, leaving a minimum of residual waste. The end goal is to retain the value of materials and resources indefinitely, with no residual waste at all.

The circular economy encourages buildings to be designed, built, operated, repurposed and/or deconstructed in a manner that minimises waste, makes disassembly easier and buildings more adaptable. In order to implement circular economy principles most effectively, it is helpful to explore high-level strategic opportunities as early in the development process as possible. The Greater London Authority (GLA) *Circular Economy Statement Guidance*[53] provides guidance for developers in London to implement the principles of the circular economy in their developments. The key principles are applicable to car park developments, including the definition of a strategic approach to the circular economy which is most appropriate to a project. Some of the different strategic approaches that can be adopted are set out in Figures 8.2 and 8.3.

Figure 8.2: Decision tree: design approaches for existing structures

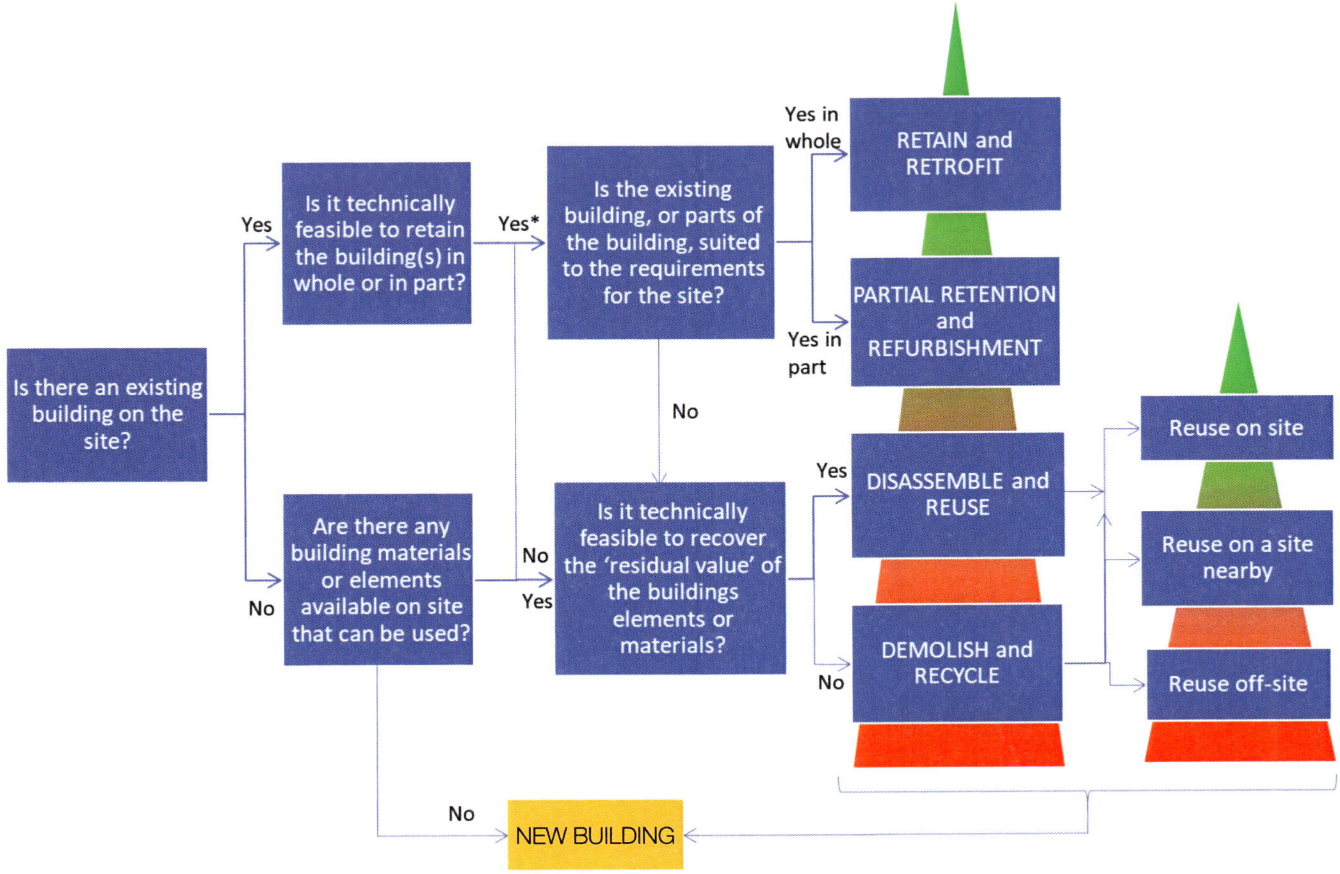

Figure 8.3: Decision tree: design approaches for new structures

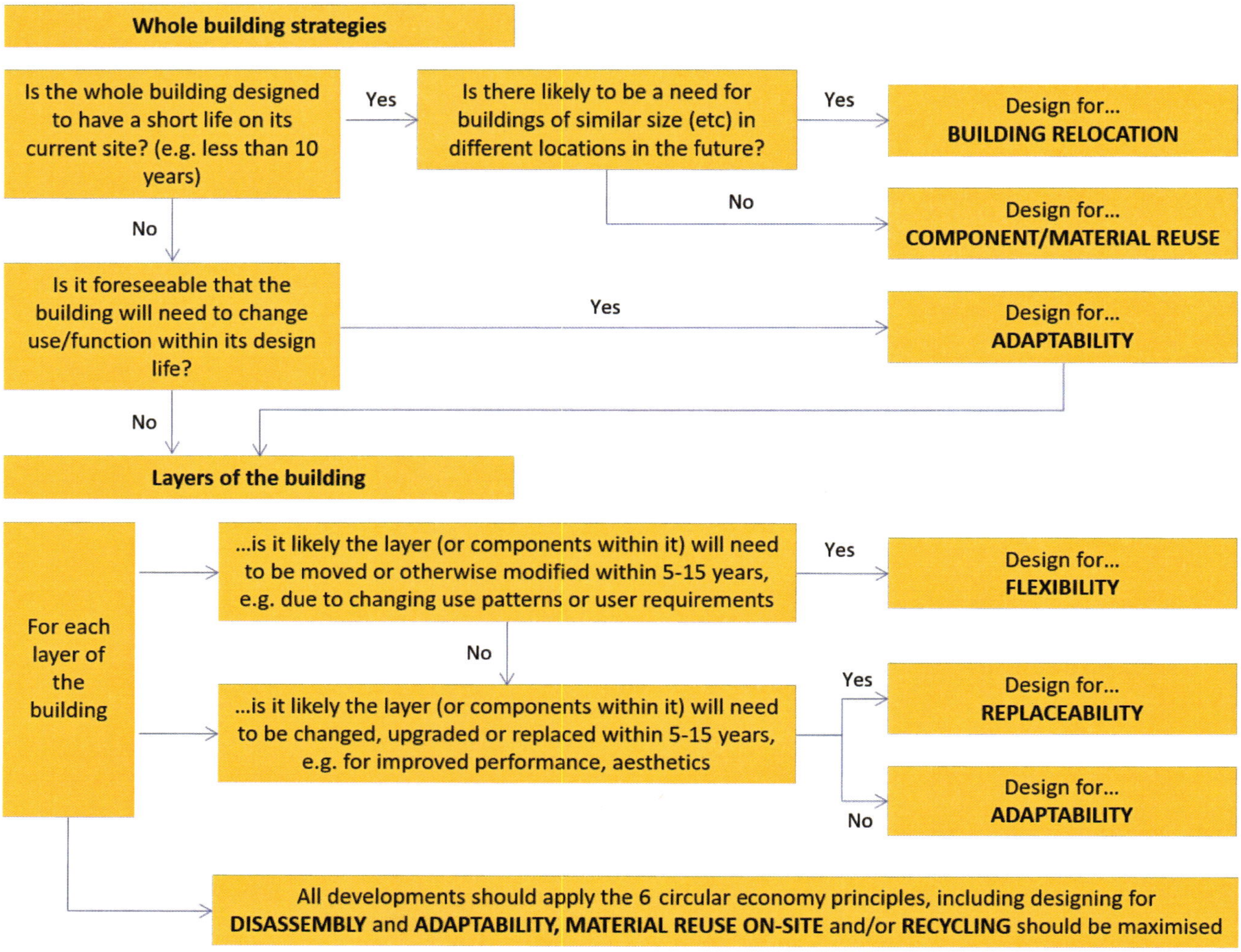

8.3.1 Long life expectancy

For a car park expected to function for many years, the strategic approach may be a combination of designing for longevity and adaptability. This should ensure that the materials are sufficiently robust to last the duration of the life of the development but may be readily adapted to a different building use type within the life of the development. In order to ensure that the development is adaptable for future uses, it is important to consider which uses which may potentially be required in future and how the structure may be adapted for these. In order to allow the structure to be retained and adapted for commercial or residential uses, generous floor-to-floor heights should be provided (Case study 1). The configuration of cores, ramps and columns should be considered, along with how other building elements (e.g., facades) could be added in future. Scenario modelling may be carried out to assess how the development could be adapted for various other uses.

8.3.2 Short life expectancy

If the expected life of the development is short, the strategic approach may be a combination of designing for reusability and recoverability. This should ensure that materials can be readily disassembled at the end of life, preferably for reuse in other projects (i.e., where materials with the same specifications and dimensions can be directly reused) or recovered for other projects (i.e., where materials can be reclaimed and remanufactured to suit different specifications and dimensions). When a development is expected to have a short life, lean design and construction principles should be implemented. The concept of lean design and construction involves minimising the amount of new construction materials, with the aim of decreasing time, effort and waste of materials. For developments which are designed for disassembly, this concept is particularly important, and must be considered in order to avoid over-specification of structural materials.

Case study 1: Parkade at the Platform Innovation Centre, Calgary, Canada

The Platform Innovation Centre (Figure 8.4) features a multi-storey car park (Parkade) with more than 500 vehicle parking spaces adjacent to the 4,645m^2 Innovation Centre. A key feature of the development is that generous floor-to-floor heights were provided so that there is sufficient space for the structure to be adapted for other uses in the future. The floor-to-floor heights allow for the additional ductwork, servicing and floor and ceiling finishes that would be required to adapt the structure for other purposes; for example for residential or commercial use. This approach maximises the potential life of the structure and safeguards against future redundancy, such as a potential decrease in demand for car parking.

Figure 8.4: Parkade at the Platform Innovation Centre

8.3.3 Design for disassembly

All multi-storey and underground car park developments should be designed for ease of disassembly. At the conceptual design stage this should include considering a modular frame approach, and designing to avoid composite, adhesive and chemically-bonded materials as far as possible, in favour of mechanically-bonded materials. It is acknowledged that many elements of car park structures do need continuity and to minimise joints and cracks (for water leakage, fire etc.) so opportunities for mechanically-bonded materials may be limited. Throughout the detailed design and construction stages it will be important to develop, and then maintain, a 'material passport' detailing the material specifications and quantities for the development, to assist at the disassembly stage.

8.4 Embodied carbon

The structural engineering strategy, including column grid spacing, structural frame and floor slab material selection, has a significant impact on the overall embodied carbon of a building. Material quantities should be minimised as far as is practical through enhanced material efficiency e.g., with consistent and efficient structural grids. A whole-life carbon (WLC) assessment should be carried out to appraise conceptual design options and to assist with optimising the structural engineering strategy.

The outputs of a WLC assessment for a conceptual design appraisal of a notional building are shown in Table 8.1. This example accounts for the embodied carbon associated with the structural frame, floor slab and foundations, for three structural frame and floor slab material options, for either a 3-storey and 6-storey building using either a 7m x 7m, or 7m x 15m grid spacing†. The table shows that the embodied carbon associated with the three different material options decreases with shorter floor spans. In order to reduce embodied carbon, floor spans should be reduced, with columns located within the parking bay separations.

Table 8.1: Lifecycle assessments for a conceptual design appraisal (notional building)

Embodied carbon (Modules A1–A5)	Option 1: Non-composite rolled steel beams with precast concrete slab	Option 2: Composite rolled steel with *in situ* concrete slab on metal decking	Option 3: Composite cell beams with metal decking
3-storey, 7m x 15m grid spacing	313	256	357
3-storey, 7m x 7m grid spacing	186	207	236
6-storey, 7m x 15m grid spacing	321	250	354
6-storey, 7m x 7m grid spacing	174	202	229

A building's substructure has a significant amount of embodied carbon associated with it; typically comparatively more than is associated with the superstructure. The requirement for basement levels of car parking (or other uses) should be determined at the project briefing stage and ideally basement levels should be designed out of the project. The weight of superstructure should be reduced as far as is practical, to decrease the size of the basement/foundations further. This includes limiting the building height and number of stories to an optimum level for building weight and embodied carbon — which is likely to be around 4-storeys high[54]. The optimum building height should be appraised for specific projects using WLC assessment.

The carbon intensity of the selected materials should be reduced as far as possible, with a preference for materials which are responsibly and locally sourced. In London, the GLA's guidance on WLC suggests a 96% recycling rate for construction waste arising, and 97% recycled content of materials for the building. However, the use of current recycled and lower-carbon structural material alternatives such as secondary steel (recycled-electric arc furnace (EAF)) and secondary cementitious materials (SCMs) may be seen as attractive 'easy wins' to drive reductions in an individual project's carbon emissions – but their supply is finite and locked into existing supply chains. These materials are already used in full by the supply chain, and so high-carbon materials currently supply the remaining demand. Furthermore, using structural materials with a higher recycled content may not always be the lowest carbon option, especially when the material is imported. For this reason, a WLC assessment should be undertaken which is specific to the project and considers the potential product sourcing, manufacturing and transport of materials.

Concrete is one of the primary contributors to a structure's embodied carbon. The amount of embodied carbon associated with concrete varies depending on the cement content which, in turn, impacts the strength of the concrete. A lower cement content reduces the strength and therefore, a greater amount of concrete may be required which could result in higher embodied carbon. There needs to be a balance between the amount of cement and the concrete strength to minimise the embodied carbon[55]. Similar to steel, recycled material can be added to concrete but the issue of limited recycled content remains.

† Car park column grid spacing is determined by bay dimensions, which are typically 2.6m x 5m. The drive aisle is typically 6m wide for one-way flow and 6.95m for two-way flow. The overall 'bin width' for one-way flow is therefore typically 16m, and to avoid the need for internal columns, the structure supporting the floor needs to span 16m. Alternatively, with columns positioned within the bay separations, shorter floor spans of approx. 7.8m may be used, provided columns are set back from the drive aisles in accordance with Fig. 5.5. However, for ease of parking, internal columns are not preferred as they reduce the dynamic efficiency. The model is based on options of a notional 3-storey and 6-storey building with a footprint of 800m^2, assumes deep (piled) foundations, and that the ground model consists of 2m of made ground followed by stiff clay.

The use of current recycled and lower-carbon structural material alternatives such as secondary steel (recycled-EAF) and SCMs such as ground granulated blast-furnace slag (GGBS), should be considered, along with new and innovative materials and design solutions.

There are design implications associated with structural material alternatives. These should be reviewed and accounted for in the project programme and risk assessment, as appropriate. For example, for concrete using GGBS, the setting time may be extended by approx. 30 minutes when compared with traditional concrete, and this may be extended at low temperatures and/or with high proportions of GGBS. This extended setting time may be advantageous in that the concrete will remain workable for longer and there will be a reduced risk of cold joints. Timber, although not widely used for multi-storey car parks at present, should be considered with caution as there are fire safety implications that must be addressed — particularly with engineered timber (e.g., glulam and cross-laminated timber (CLT)), which pose the risk of fuelling a fire as the adhesives between the layers can melt and the fire can burn the timber from the inside out.

8.5 Operational carbon

Multi-storey and underground car park developments will need to maximise opportunities to reduce operational carbon emissions, generate renewable energy on-site, and to meet the present and future demand of charging from electric vehicles. The following are considerations for structural engineering at the conceptual design stage.

8.5.1 Energy strategy

As part of a planning requirement, any development will require the submission of an energy strategy. A good practice guide is the GLA's *Energy Planning Guidance*, which sets out the 'Be Lean, Be Clean, Be Green, Be Seen' *approach*[34].

8.5.2 Electric vehicle (EV) charging

The UK Government has now confirmed that the sale of new petrol, diesel and hybrid cars will be ended by 2030, with all new cars and vans being fully zero tailpipe emissions from 2035[56]. Both current and future demand for EV charging should be taken into account; this is likely to include the installation of EV chargers along with electrical infrastructure to enable their installation to an increased proportion of car parking spaces in the future.

8.5.3 Renewables

A key opportunity to implement renewable energy generation as part of car park developments is with roof-mounted solar photovoltaics (PV) on multi-storey car parks. PV panel areas should be maximised to generate as much renewable electricity on-site as possible, to contribute to meet the electricity demands of the car park, including EV charging (and may contribute to energy demand in associated buildings). PV panels may also be mounted on top of green roofs. There will be constraints to consider, including roof plant space requirements, height of structure, massing and visual impact and adjacency to nearby buildings. The loads imposed by PV panels and green roofs should be considered in structural design.

8.5.4 Energy storage

Energy storage, for example with batteries, allows system flexibility in response to specific energy demands (e.g., EV charging) and allows for the stockpile of renewable energy generated on-site for times of demand, or for drawing electricity from the grid at times when off-site renewables are generating electricity and the carbon intensity is lower, to use during peak demand when carbon intensity from the grid may be higher.

At present, it is understood that the cost of battery storage at a development-scale is significant and the payback times on such technologies for developers are therefore relatively long. Batteries and other technologies to convert and store electricity are currently being explored at the regional and grid-scale. In the future it is likely that the business case for development-scale storage will become more favourable with storage costs reducing as the technology matures, along with potential increases in energy prices and the transition to time-of-use tariffs.

8.6 Greening and biodiversity

As mentioned in Section 8.1, the UK Government published its *25 Year Environment Plan* in 2018, which sets out what will be done to improve the environment in a generation. The Environment Act 2021 includes a requirement for 'biodiversity net gain' for developments, which requires a 10% increase in biodiversity to be demonstrated on or near development sites[44].

8.6.1 Sustainable drainage systems (SuDS)

SuDS provide alternative drainage to natural waterways rather than using a traditional piped sewer network. Well-designed SuDS will encourage new habitats and enhance the local biodiversity including plants, amphibians, birds and mammals. They help reduce the risk of flooding by reducing flow rates and with increased instances in the UK, SuDS could help lessen the impacts of severe flooding. The Construction Industry Research and Information Association (CIRIA) *SuDS Manual* provides a comprehensive guide to SuDS and further detail on how systems may be integrated in project sites[57].

8.6.2 Rainwater harvesting

The collection of rainwater from roofs for reuse on-site is a vital component in delivering a building which is resilient and minimises its reliance on the mains potable water supply. Water may be harvested for use in adjacent facilities, for example flushing WCs, or for landscaping irrigation (including green roofs and walls).

8.6.3 Green roofs and walls

Green roofs can vary in their design and function, typically being classed as either 'extensive green roofs' (a green roof with substrate depth less than 200mm, supporting sedum, moss or grass) or 'intensive green roofs' (a green roof with a greater substrate depth, often 1m or greater, supporting sedum, moss, grass, shrubs or trees). For projects in urban areas, with limited availability of land for ground-level landscaping, green roofs are likely to be necessary to meet the 'biodiversity net gain' requirements. Green roofs may be designed to include recreational and amenity spaces, along with providing wildlife habitats for birds and pollinators such as bees, with specially selected native plants. Extensive green roofs may also used in conjunction with solar photovoltaic (PV) panels as an integrated 'biosolar roof'. The green roof can help to regulate the temperature of the PV panels, thereby increasing the efficiency of the panels and increasing electricity generation. Green walls, like green roofs, provide more habitats for animals and they help to reduce the cooling and heating demand of a building (Case study 2).

Case study 2: Park 'n' Play car park, Copenhagen, Denmark

The 'Park 'n' Play' car park in Copenhagen was designed to encourage people to use the car park roof by including a play area equipped with swings, ball cages, and jungle gyms. The facades of the car park incorporate a series of plant boxes, distributing the greenery across the entire facade (Figure 8.5).

Figure 8.5: 'Green walls': Copenhagen's Park 'n' Play car park

8.7 Certification and verification

There are a number of sustainability assessment and rating methodologies which are applicable to building and infrastructure projects. Table 8.2 provides an overview of key methodologies available for projects in the UK.

Some of these methodologies are applicable only to occupied buildings and may be appropriate if a multi-storey or underground car park forms part of a wider development, associated with an occupied building/s. Other methodologies are applicable for infrastructure projects and may be used for standalone car parks.

8.7.1 BREEAM Infrastructure

Formerly known as 'CEEQUAL', BREEAM Infrastructure[58] is a BRE-owned sustainability assessment and rating methodology for the assessment of all types of civil engineering, infrastructure, landscaping and public realm projects and contracts. The BREEAM Infrastructure Projects scheme is for civil engineering, infrastructure,

Table 8.2: Summary of key sustainability assessment and rating methodologies

Scheme	BREEAM	BREEAM Infrastructure	LEED	Living Building Challenge	WELL Building Standard
Latest scheme version/s	BREEAM UK New Construction 2018, BREEAM UK Refurbishment & Fit-Out 2014	BREEAM Infrastructure Projects UK & Ireland, Version 6	LEED v4.1	Living Building Challenge v4.0	WELL Building Standard v2
Rating Authority	BRE Global	BRE Global	US Green Building Council (administered by Green Business Certification Inc.)	International Living Future Institute	International WELL Building Institute (administered by Green Business Certification Inc.)
Originating country	UK	UK	USA	USA	USA
Scheme coverage	Global	Global	Global	Global	Global
Eligible project type(s)	Occupied buildings	Infrastructure projects	Occupied buildings	Occupied buildings	Occupied buildings and infrastructure projects
Content	Covers sustainability topics, including: health and wellbeing, energy, transport, water, materials, waste, land-use and pollution.	Covers sustainability topics, including: management, resilience, communities and stakeholders, land use and ecology, landscape and historic environment, pollution, resources and transport.	Covers sustainability topics, including: transport and location, water, energy, pollution, materials and indoor environmental quality.	Covers sustainability topics, including: water, energy, health and happiness, materials and equity.	Covers health and wellbeing topics, including: air, water, nourishment, light, movement, thermal comfort, sound, materials, mind and community.

landscaping and public realm works, and may be applied to new car park development. BREEAM Infrastructure Version 6 can be used to assess and rate the sustainability performance of infrastructure projects at the Strategy, Design (or Interim Design) and Construction stages. Several car park case studies are available from BRE[59].

8.7.2 Living Building Challenge

The Living Building Challenge (LBC)[60] aims to transform how we think about every single act of design and construction, as an opportunity to positively impact the greater community of life and the cultural fabric of our human communities. LBC 4.0 is a holistic standard, pulling together the most progressive thinking from the worlds of architecture, engineering, planning, interiors, landscape design, and policy. The LBC is versatile and applies to different project scopes, or 'typologies'. The four typologies are 'new building', 'existing building', 'interiors' and 'landscape or infrastructure'.

8.8 Operational performance verification

The performance of a net zero carbon car park will be determined once the construction is completed and the final upfront embodied carbon associated with the construction has been accounted for. It continues in operation, as operational carbon is emitted as a result of energy consumption *and* from the *embodied carbon* associated with management, maintenance and replacement while the car park is in operation.

In buildings, there are methods available to improve the accuracy of energy consumption predictions, verify design of the building and ensure a smooth transition from design and construction, through to operation. The Post Occupancy Evaluation (POE)[61] and Soft Landings Framework[62] are examples of methods which aim to achieve this, while recognising that the ongoing business benefits, maintenance, and operational cost of a building during its lifecycle far outweigh the original capital cost of construction. The need to optimise in-use performance is recognised and sought through early engagement with the operators and end users in the design process, and by learning from previous experience.

8.9 Offsetting

Carbon offsetting is the last stage in the net zero carbon hierarchy and aims to capture carbon from the atmosphere or to minimise emissions elsewhere to achieve net zero. Although carbon offsetting is vital in achieving net zero, it should only be used to compensate unavoidable emissions, and in conjunction with other net zero principles such as 'build less' and 'specify low carbon materials and technologies'. In the future, the need for carbon offsetting should greatly reduce as more low carbon materials and technologies enter the market. The Institution of Structural Engineers has published a short guide to carbon offsetting, including the key considerations for engineers looking to offset the carbon emissions from projects[63].

9 Safety and security

9.1 Introduction

This chapter reviews the various safety-related considerations associated with multi-storey and underground car parks. Equally important, and indeed key to being able to facilitate the design process, is the comprehensive understanding and appreciation of obligations which will be faced by the car park operator. Without taking this into consideration the designer cannot reasonably fulfil the obligation to design an appropriate facility to meet the client's brief.

The design process associated with safety and security aspects should follow the principle of 'avoid, control or accept'. In other words, the designer should assess each safety or security risk based on its own merit and in the context of the building being designed, and should either eradicate the risk, mitigate it or accept that the risk will remain, and ensure that the residual risk is correctly highlighted and communicated to the relevant party. This process is collaborative and should involve the entire design team as well as the client's team, and ideally would take the form of a workshop where risk is discussed openly.

There can be benefits in discussing security arrangements with the client and the Police Designing Out Crime Officer (DOCO) (Section 16.3). The level of security and principles for the provisions required should be discussed and agreed at an early stage of the conceptual design.

9.2 Obligations on designers

The obligations on a designer of a multi-storey or underground car park are many and varied. Principally, and at the source of all obligations, designers must adhere to the statute of law.

For the design of a UK car park, as with many other buildings, compliance is required with the following:

- The Town and Country Planning Act[64]
- The Building Regulations[10]
- The Construction (Design and Management) Regulations 2015[3]

9.2.1 The Town and Country Planning Act

This covers the legal process of determining whether proposed building works should be permitted, and is the responsibility of local planning authority. This process will consider many aspects such as the location, size, massing and appearance of the proposals together with any specific local authority policies. The local authority will request advice from its internal departments as appropriate for the development in question e.g., highways, environmental, conservation, etc.

Some safety and security matters are considered within the Act e.g., whether or not the building will increase risk to the existing safety and security of the area. Another example is the safety of the highway network and associated components and how this, and pedestrians, are affected by the proposal.

Early consultations are always recommended, as planning applications can be refused on the grounds of unsafe highways access.

9.2.2 The Building Regulations

The Building Regulations are the main set of statutory approvals that UK-based designers need to comply with.

It is very important that consideration be given not just to those utilising the building via entering the car park, but also for people who may be affected indirectly by the building and its use e.g., pedestrians walking adjacent to the building to access other buildings, or even those utilising buildings nearby which may be affected e.g., in the event of a fire.

A series of Approved Documents[65] provides guidance on how to comply with the Building Regulations. However, there is no obligation to adopt the solutions presented within these, as the Building Regulations can be satisfied in other ways. For example, British Standards can be used by designers to demonstrate a different approach to provide compliance. One example is the use of *BS 9999: Fire safety in the design, management and use of buildings*[31], which can provide some alternative methods of obtaining compliance with statute, as opposed to the adoption of Approved Document B[66]—the Approved Document relating to fire.

In practice, the Approved Documents are the most common method adopted to ensure compliance for common building situations.

If at any stage, a car park is found to be unsafe and at risk of collapse or danger to the public, it can be brought to the attention of the local authority. If the owner/occupier fails to act, the local authority can issue an enforcement notice to have the building closed, or alternatively can undertake the necessary works to make the building safe (temporary propping and/or repairs) and recover the cost from the owner/occupier.

9.2.3 The Construction (Design and Management) Regulations 2015

The CDM 2015 Regulations are the most significant set of regulations for designers to consider when designing a car park. They are intended to ensure that all health and safety and welfare issues are properly considered and addressed during a project's development, so that the risk of harm to those that construct, use, maintain and decommission is eliminated (and where this is not possible, reduced or highlighted).

Compliance with the regulations ensures that designers consider all aspects of the risks associated with construction and maintenance of buildings, from the outset of the design process. Specific consideration should be given by the designer to design out the risk or, where this not possible, mitigate the risk and above all ensure that the resultant risk is clearly communicated to the relevant party.

9.3 Obligations on owners/operators

The primary obligations on owners and operators of UK car parks with regard to safety and security are defined within two pieces of legislation: the Occupiers Liability Act[67] and the Health and Safety at Work etc. Act[68].

Common law also applies — in so much as to put an obligation on the owner/operator to ensure the building is safe for use by those visiting the building.

9.3.1 Occupiers Liability Act

The Occupiers Liability Act 1984 extends the common duty of care on the owner/operator of the property to trespassers as well as visitors. If the owner/operator knows there is a risk, the risk must be mitigated otherwise the owner/operator is liable for any consequences.

In effect, the owner/operator should not permit or in any way present a situation where a trespasser may access the property and be subjected to a risk that exists within or around the building. It is therefore reasonable to summarise that the owner/operator should ensure at all times that the building is not presenting any risk to any persons at any time which had not been reasonably removed or mitigated to an acceptable level.

In order to ascertain whether acceptable measures have been taken to reduce risk, a risk assessment process should be followed, documented and, wherever practicable, should be influenced by current Building Regulations.

There is no general requirement to retrospectively upgrade a facility or car park to adhere to current Building Regulations, although specific exceptions may from time to time come into force. However, the obligation to ensure that the building is safe remains, and therefore when measures are not taken to upgrade to current Building Regulations requirements, the decision should be based on a considered and informed process of risk-assessed design (Section 9.5.1). Justification will no doubt be required as to how the building could be deemed safe for use despite not being compliant with current Building Regulations.

9.3.2 Health and Safety at Work etc. Act

The Health and Safety at Work etc. Act 1974 is the primary piece of legislation covering occupational health and safety in the UK. It is sometimes referred to as HSWA, the HSW Act, the 1974 Act or HASAWA.

It sets out the general duties which:

- Employers have towards employees and members of the public
- Employees have to themselves and to each other
- Certain self-employed persons have towards themselves and others

The designer should consider the fact that any car park is almost certainly going to be a place of work, and therefore the owner/operator will have an obligation to adhere to the Health and Safety at Work etc. Act.

While the Act itself is extensive, ultimately, a conclusion can be drawn that employers have an obligation to their employees to provide a safe place of work. If, as is likely, a car park has an attendant or a cleaner etc., then it is a place of work and falls under the Act.

Depending on the proposed method of operating the car park, it may be appropriate (and a requirement) to provide welfare facilities for staff whether they be fully based in the car park e.g., parking attendants, or whether they are occasional visitors e.g., cleaning teams.

The method of operating the car park must be recognised during the design, so as to ensure that a safe place of work is provided, and so the owner/operator is able to comply with their obligations under the Health and Safety at Work etc. Act.

The operator's ability to correctly and adequately maintain the car park requires consideration because if deterioration of the car park occurs and is not arrested, it may become an unsafe environment even if this was not the case previously.

9.4 Primary risks

The safety and security risks to visitors are varied and numerous. The most common are outlined in the following subsections.

Ultimately, the designer should do all they reasonably can to ensure they are informed of the risks during the design stage so that these can be removed or reduced. If all else fails, the risks must at least be understood, documented and highlighted.

9.4.1 Slips, trips and falls

Consideration should be given to the finishes of floor surfaces throughout the car park — with specific attention being paid to the fact that the floors may become slippery (e.g., the top deck of a car park may suffer from flowing water, especially on inclined decks).

The fact that vehicles also traverse areas where pedestrians walk should be considered in terms of degradation of the surface or delamination of any waterproof membrane in place, which may lead to increased risk of slips and trips.

The most common areas for trips to occur is in the degradation of the wearing course of the main car park decked area (e.g., potholes) and at transitions between differing zones e.g., stair cores and parking areas. Careful consideration should be given to the rated degradation of the specified wearing course — whether that be a sand/cement screed or mass concrete, or a form of macadam. Consideration should also be given to the methods available to mitigate risk of trips at junctions of differing floor areas e.g., utilisation of movement joints.

A floor risk assessment should be undertaken routinely by owners and operators, but consideration should be given by the designer to ensure that the owner/operator is able to maintain the floor to an acceptable standard.

Utilisation of pendulum slip testers (or similar appropriate), and the desired readings to gauge and define the level of slip risk presented, is necessary.

The enhanced slip risk is often due to the mixed nature of gradients which pedestrians are subjected to within the building, and this should be considered during the design stage. One way of mitigating this is the utilisation of demarcation deck signage to aid pedestrians when they are walking through or around the car park.

Trips can often occur in areas where there is a change in level that has not been properly graded out e.g., where the transitions to vehicle ramps meet the drive aisles and where there are raised kerbs.

Falls from height are a particular risk in car parks, especially due to the fact that car parks are often open-sided with limited edge protection. There is also a risk of falling within the building between levels of the car park, especially on car parks which are designed with a split-level arrangement.

Although slips, trips and falls are risks commonly associated with the physical condition of the surface of the car park, the importance and relevance of lighting should not be underestimated. Lighting should be consistent in terms of coverage and lux levels within defined areas, and sufficient to enable users to be able to recognise changes in level.

9.4.2 Suicide

Suicide risk itself is not specifically covered within the Building Regulations or statute other than to say that the building should be safe for use by all and that there is no clear exclusion of this rule to those who are intent on committing suicide.

It should be noted that the recommendations within the Approved Documents relating to edge protection are based on the requirements to prevent accidental falling and not necessarily to prevent the intentional act of jumping from the car park. Further guidance is provided in Chapter 16.

9.4.3 Theft

Theft of property either from a vehicle, from a person, or theft of the vehicle itself, is a risk within car parks as it is elsewhere on the highway network.

Arguably, the risk is increased within a car park due to the sheer density of property within the car park at any one time, and also due to the fact that car parks can present areas of limited visibility which could be seen as attractive to criminals. Work has been carried out in recent years to combat this e.g., the British Parking Association's Safer Parking Schemes[17]: 'Park Mark®' and 'Park Mark Plus®' — which strive to minimise risk via design and operational measures and subsequently make the building a safer place to park.

9.4.4 Personal attack

As with theft, attacks on individuals is a risk within car parks as it is elsewhere in society. However, the perceived risk is greater in a car park environment due to the fact that they are often environments with limited visibility. Views into the car park may be shielded by facade treatments or similar, and especially during hours of darkness, the car park may be relatively empty. As with theft, considered design and adhering to schemes such as Park Mark® should aim to minimise this risk.

9.4.5 Drug abuse and anti-social behaviour

Typical car park conditions (muted lighting, less than optimal visibility/sightlines, partial cover from inclement weather etc.) can attract anti-social behaviour (e.g., urination and vandalism) as well as drug-related activity.

9.4.6 Pedestrian-vehicle conflict

Pedestrians and vehicles need to circulate the car park both horizontally and vertically, in relatively confined spaces. Occasional conflict between the two is inevitable.

This risk is particularly apparent when pedestrians are entering/exiting their vehicle or crossing areas in which vehicles are passing.

Limited visibility associated with level changes and partitions etc. exacerbates this risk within some car park layouts.

9.4.7 Containment of vehicles

It is important to ensure that sufficient edge and key component protection is in place to mitigate the risk of impact from a vehicle or the risk of a vehicle being able to fall from the car park.

Ultimately there is a need to have vehicle impact barriers around the perimeter of the decks and at changes in level to prevent vehicles from moving beyond the deck edge.

Driver error can result in significant force or repeated low-force impacts on barriers and cladding systems which can lead to loosening or displacement; putting persons outside the car park at risk from falling debris, and even falling vehicles.

Significant relevant guidance and requirements (including detail on vehicle restraint systems) is provided in Chapter 12.

9.4.8 Terrorism

Due to multi-storey and underground car parks generally being located in densely populated areas, often in close proximity to significant buildings e.g., shopping centres, they are at increased risk from terrorist attack.

At airports or near prominent public buildings (e.g., courthouses) where the car park can be used to discharge weapons, screening within the car park is required. Car parks below statement or control buildings can give access to the supporting structure and require additional levels of security.

9.4.9 Fire

The risk of fire needs to be considered very carefully at every stage of the building's life and use. This is discussed in both Chapter 7 (in relation to conceptual design) and Chapter 14 (for detailed design).

9.5 Designing out risk

When attempting to design out risk associated with safety or security, the same approach should be taken as with all other risks i.e., the risk should be fully understood and the factors which affect the level of risk should be considered in order that safe design solutions may be adopted.

One key area of context when understanding contributing factors to risk is the location and more specifically the user group. For example, a car park located in an out of town retail park is likely to have an entirely different user group from a car park located within a large hospital complex, especially if the hospital includes a mental health facility. There are clear and significant differences to the potential for some risks (e.g., suicide) between these scenarios.

Other potential differences in user groups may be the duration each visitor will stay, as this will increase turnover of cars and pedestrians which will in turn increase the number of potential conflicts between vehicles and pedestrians. An example of this may be a train station car park, which may typically be busy during the commuting hours but quiet at all other periods, versus a convenience shop car park, which may have a very high turnover of vehicles throughout the day.

The following two subsections suggest how designing out risk may be undertaken.

9.5.1 Risk-assessed approach

The ultimate obligation to ensure that the building is safe for use may result in the design of the car park needing to go above and beyond the design solutions described in the Approved Documents. The designer therefore cannot rely solely on following these to achieve statutory compliance.

Where the car park will present risks beyond the basic principles adopted in the Approved Documents, the designer should assess the risk on its own merits and in the specific context of the car park being designed — ideally utilising a method which requires the designer to assess the likelihood of the risk occurring and the severity of the outcome, in order to determine the risk rating.

Adopting such a process should lead to the designer carefully considering the safety and security risk, and will provide an audit trail (Figure 9.1 at the end of this chapter serves as an example risk assessment). Such process is also linked directly to the CDM 2015 Regulations where the designer is obliged to remove the risk if they can, and if they cannot, mitigate the risk to the point where any risk remaining is articulated to the next duty holder of the project, which may well be the contractor, but will ultimately be the owner/operator.

9.5.2 Park Mark®

The Safer Parking Scheme from the British Parking Association is a national standard for UK car parks which has low-crime measures in place to ensure safety of people and vehicles.

A Park Mark® is awarded to car parks that achieve robust safety standards. The distinctive signage (Figure 9.2) is designed to help drivers find car parks where they can leave their vehicle knowing that the environment has been deemed 'safer' by the accreditation scheme. According to Park Mark®, vehicle-related crime has been falling since its peak in the mid-1990s.

Figure 9.2: Park Mark® signage

Ultimately the Park Mark® scheme aims to encourage car park owners, operators and designers to work towards a standard which includes aspects such as lighting and access control and CCTV, in order to reduce the risks associated with safety and security, with the ultimate aim to provide safe and secure car park facilities.

The utilisation of this standard at the design stage — i.e., making the adherence to the Park Mark® award criteria an employer's requirement, and therefore a part of the designer's brief — is often used as a way to ensure that the designers are deploying design methods and options with a view to reducing risks associated with personal safety and security.

Park Mark Plus® is a relatively new and more comprehensive assessment when compared to the standard Park Mark® award.

To achieve Park Mark Plus® status, car parks must demonstrate:

- Low levels of crime and measures in place to ensure the safety of people and vehicles
- The ability to meet Disabled Parking Accreditation[69] criteria for accessible parking bays
- Structural integrity
- Environmental credentials
- Good management and operations
- Exemplary customer experience

Park Mark Plus® is only awarded to facilities that meet the mandatory requirements of Park Mark® in addition to the requirements of accessibility for disabled customers in line with Disabled Parking Accreditation standards.

9.6 Potential design solutions

9.6.1 Slips, trips and falls

Where possible, the designer should adopt a consistent approach to floor finishes. Consideration should be given to potential degradation from exposure to natural elements and to mechanical wear from pedestrians and vehicles.

Areas with potential additional risks (e.g., stair cores adjacent to driving decks) should have interfaces between the zones which do not present a trip hazard.

Where deck coatings are utilised, the inevitable wearing of these should be considered — bearing in mind that coatings will often wear in an uneven pattern, with wear accelerated in areas of vehicle turning or increased football.

If deck signage or deck demarcation is utilised, consideration should be given to the likelihood of this leading to a localised increase in slip risk due to the 'levelling' of the deck surface at those locations, unless the demarcation is itself textured.

The minimum height for pedestrian guarding is 1,100mm measured from the top of the wearing course to the top of the edge protection. There should also be no gaps which would allow a 100mm diameter sphere to pass through, and the guarding should not be climbable in any car park which may be accessed by children.

While this is the minimum standard, it does not necessarily represent the standard or extent required to adequately remove or mitigate the risk of falling from the building or jumping across voids in stair cores, and the designer should bear this in mind when conducting the risk assessment.

The perceived risk of falling from the building can be mitigated by using solid panels as opposed to mesh or glass and by using materials or systems which are rigid rather than flexible.

9.6.2 Vehicle impact

The risk of vehicles falling from the car park is considered in Chapter 12.

9.6.3 Criminal activity

The mitigation of risk of criminal activity such as theft, drug use and other anti-social behaviour in a car park is much the same as in other buildings or public spaces.

The utilisation of techniques to counteract the potential features which can encourage such behaviour or activity should be encouraged. For example, one feature of a car park likely to appeal to a person intent on anti-social behaviour could be the relatively enclosed nature of the building, which significantly limits the chance of that person being seen committing the act. A way of counteracting this would be utilisation of an open layout which does not have areas of limited vision. Another tool might be the deployment of CCTV.

Good lighting is another technique which has been adopted. Light fixtures and fittings should incorporate vandal-resistant features, with cables and wiring securely enclosed (Figure 9.3).

Access control systems (e.g., barriers) can be utilised, so the building is not perceived as being entirely open and accessible to all. The utilisation of pedestrian access control increases public perception of safety and security and further dissuades anti-social behaviour.

9.6.4 Pedestrian-vehicle conflict

Using colour coding or deck demarcation can be a positive way to mitigate against risk of conflict as it promotes segregation of pedestrians from vehicles, but no amount of visual segregation will remove the risk completely as pedestrians and cars will often share the same spaces.

Increased and unimpeded visibility across the whole car park can reduce this risk further e.g., using a mesh instead of a solid wall.

Figure 9.3: Effective illumination

Figure 9.1: Example risk assessment

Design Risk Assessment

This Quantitative Risk Assessment produces a numerical representation of the expected frequency, and extent of a specified level of exposure or harm, to specified people on and off-site, from a specified activity.

Designers should aim to eliminate hazards within their designs; consideration should be given to the construction, use, maintenance and eventual demolition of all structures and the hazards associated therein.

Stripe Consulting uses a Quantitative Risk Assessment to; highlight specific tasks, consider existing control measures, establish a risk rating per task, consider additional control measures and establish a residual risk rating.

Hazard – is the potential to cause harm e.g., collision with vehicle

Control Measures – are the actions taken to prevent harm as a result of the said hazard, such as traffic segregation, implementing on-site speed limits, etc. Control Measures include such areas as training, supervision, instruction, information, safe systems of work, proper maintenance procedures, as well as physical measures such as guard rails (to prevent falls), barriers (to prevent access to hazardous areas), guarding of machinery, etc.

Risk – is the likelihood that harm will occur, after the control measures have been carried out.

The equation for establishing risk ratings is Likelihood of Injury (1-5) (L) x Severity of Injury (1-5) (S) = Risk Rating (1-25)

		Likelihood				
		1	2	3	4	5
Severity	1	1	2	3	4	5
	2	2	4	6	8	10
	3	3	6	9	12	15
	4	4	8	12	16	20
	5	5	10	15	20	25

Likelihood x Severity = Risk Rating			
Likelihood	**Severity**	**Rating**	**Action**
5=inevitable	5=Fatality or major injury	**Over 20**	**Intolerable risk** – HIGH immediate action required
4=highly likely	4=Serious injury resulting in days lost	**15-19**	**Substantial risk** – HIGH control measures to be put in place
3=possible	3=Possible hospital treatment	**10-14.**	**Moderate risk** – MEDIUM control measure to be put in place
2=unlikely	2=First Aid may be required	**5-9**	**Tolerable risk** – MEDIUM control measures to be considered
1=highly unlikely	1=Unlikely to cause harm	**Below 5**	**Trivial** – LOW implement action if indicated

Low (1 - 4) **Medium (5 - 14)** **High (15 +)**

Figure 9.1: *continued*

Project	
Job No.	

No.	**Designer**	**Project stage**	**Activity**	**Hazard**	**Drawing reference location no:**	**Persons at risk**			**Likelihood (L) of injury (1-5)**	**Severity (S) of injury (1-5)**	**Risk rating (L x S)**
						Operatives	**Staff**	**Public**			
1		Construction	Working adjacent to highways.	Risk of injury by moving vehicles.	Site plan	Y	Y	Y	2	4	8
2		Construction	Work with lifting and manoeuvring large scale components of deck system. Including possible use of mobile crane.	Risk of injury by falling obects.	Site plan	Y	Y	Y	1	5	5
3		Construction	Work to upper floor decks prior to installation of edge protection.	Risk of operatives falling.	Site plan	Y			3	5	15
4		Construction	Installation and de-mounting of cladding panels.	Risk of operatives falling/suffering injury while handling large panels.	Sections	Y			3	5	15
5		Construction	Works to areas of sloped deck, including overhead items e.g. installation of soffit mounted services. Non-flat surface for scaffold towers/ladders etc.	Risk of wheeled items rolling away. Risk of operatives falling due to unstable platforms.	Floor plans	Y		Y	1	4	4
6		Construction	Work to roof of cores prior to installation of edge protection.	Risk of operatives falling.	Roof plans	Y			3	5	15
7		Construction	Working adjacent to steeply sloping banks/uneven ground on northern and eastern boundaries of site.	Risk of operatives falling.	Site plan	Y			2	4	8
8		Construction	Proximity of works to live railway and risk of items/materials falling or being blown onto railway.	Risk to railway opertives and users.	Site plan	Y		Y	2	5	10
9		Construction	Construction/excavation near existing underground services.	Risk to operatives of electrocution/ damage to services.	Site plan	Y	Y		3	3	9
10		Construction	Excavation for foundations (assumption for pile caps and core foundations prior to the soil investigation results becoming available) and retaining walls.	Risk of soil not being able to bear its weight.		Y			3	5	15
11		Construction	Installation of steel reinforcement for foundations (crane lifts, positioning in excavated pits).	Risk of operatives suffering injury while handling/ installing.		Y			2	4	8
12		Construction	HV & LV electrical connections.	Risk of electrical shock.	N/A	Y	Y		3	4	12

10 Futureproofing

10.1 Introduction

Futureproofing to meet the needs of a changing society secures commercial viability of the car park over the long-term. Unless a car park has been designed with futureproofing in mind, it will likely need to be adapted, which requires compromise and can lead to loss of functionality. Futureproofing can be challenging, particularly when the service life of the car park is significantly longer than that of the vehicles using it. Most vehicles are in service for around ten years whereas car parks, if properly designed and maintained, can be operational after more than eighty years.

Evolution in vehicle design and advances in technology have been faster than previously thought possible. This is driven by the need to reduce carbon emissions, and in the UK the intention is for fossil fuel vehicles to be no longer sold after 2030 and hybrids to be no longer sold after 2035[56]. Needless to say, fossil-fuelled vehicles will still be on the roads for some time after this date. While hydrogen-powered vehicles offer a credible alternative to fully electric vehicles, the present effort is to develop a charging network to enable batteries to be the dominant source of energy.

Currently, the time taken to recharge a battery electric vehicle at fast rate is approx. 30–40 minutes which is still several times longer than refuelling with petrol or diesel. Charging time is longer if the charge rate is slower and therefore public electric refuelling requires associated short-stay parking.

10.1.1 Vehicle ownership vs. hiring/autonomous

Over a 12-month period of mixed driving, the average speed of a vehicle is around 30mph even though some of the driving may be at motorway speed. For an annual mileage of 12,000 miles the number of hours spent in motion is therefore 12,000/30 = 400 hours. This equates to just over one hour per day and for the other 23 hours the vehicle is stationary. The cost of owning a vehicle is therefore very high especially when considering the limited time spent in motion, so the future trend is likely to move more in favour of hiring to suit the needs of the journey. The Uber model of calling up a vehicle through an app is therefore attractive and when the journey is complete the vehicle returns to base or to the next customer.

Self-park vehicles have been around for a while and technological advances have enabled autonomous vehicles to be developed which can transport passengers without the need for a driver.

Both options allow vehicles to be used for longer periods and will be a more economical option than vehicle ownership. If a hire company owns the vehicles, they can be in use for over 18 hours a day before returning to base (possibly a car park) for cleaning, servicing and reenergising.

10.2 Vehicle design

Competition between motor manufacturers drives innovation, with drivers increasingly preferring higher levels of luxury and technology. This previously led to an explosion in 4x4s, regardless of whether they were the most suitable vehicle for everyday use. The ability of 4x4s to tackle most terrains has spawned the birth of sports utility vehicles (SUVs) which are more versatile for family use and their popularity has grown rapidly. In general, this has contributed to an increase in vehicle size and weight over time.

10.2.1 Means of propulsion

While hydrogen to power an internal combustion engine offers significant advantages over battery power, the present government policy is to prioritise installation of a national network of charge points which will inevitably make electricity the standard means of propulsion for cars and vans i.e., the typical vehicles that populate car parks.

The current alternatives to fossil-fuelled vehicles comprise:

- Battery Electric Vehicles (BEVs) i.e., vehicles powered solely by a battery that needs recharging
- Plug-in Hybrid Electric Vehicles (PHEVs) i.e., vehicles that have both a battery and a petrol or diesel engine, but the battery can be charged from a charge point

- Hybrid Electric Vehicles (HEVs) i.e., vehicles with both a battery and a petrol or diesel engine and relies on the engine to charge the battery
- Mild Hybrid Vehicles (MHEVs) i.e., vehicles that use a battery to power an electric motor to support a petrol or diesel engine for improved efficiency
- Fuel Cell Electric Vehicles (FCEVs) i.e., vehicles powered by hydrogen to feed a fuel cell

All of these require refuelling at some point, and battery electric vehicles and plug-in hybrid electric vehicles require electric charge points to reenergise their batteries.

Alternative greener fuels to petrol, diesel and biofuels are currently being developed that are more environmentally friendly and could potentially enable current internal combustion engine vehicles to remain in service for the foreseeable future.

10.2.2 Vehicle sizes and weights

Undoubtedly vehicles have become wider, which assists battery-powered models/variants, because the larger the battery the greater the power and hence the longer the range. While it is difficult to predict the future shape of vehicles, it is likely that their size will continue to fit into the geometry of the SDV (Section 5.2.1).

However, while the SDV is likely to continue to encompass all the vehicles previously described, the future weight of electric vehicles is subject to conjecture. The average vehicle weight has increased significantly since the loadings for car parks were first configured. At that time the average gross weight of cars was in the order of 1,500kg and the maximum gross weight 2,500kg. The average gross weight of electric vehicles is currently around 2,000kg and the maximum gross weight can be significantly more than 3,000kg. However, with advances in battery technology it is likely that battery electric vehicles will become lighter over time.

10.2.3 Autonomous vehicles

Autonomous or 'smart' vehicles embrace artificial intelligence which was first used in self-park and lane-departure systems and is well established. The goal is for vehicles to be fully autonomous, which is becoming an increasingly likely future scenario.

As autonomous vehicles will be able to park themselves, the impact on the long-term future of car parks is considerable. Potentially there could no longer be a mix of cars and pedestrians in the same building, which will change the way car parks are configured.

The Society of Automotive Engineers describe five levels of autonomy[70]:

Level 0: No Automation. The full-time performance by the human driver of all aspects of the dynamic driving task, even when enhanced by warning or intervention systems.

Level 1: Driver Assistance. The driving mode-specific execution by a driver assistance system of either steering or acceleration/deceleration using information about the driving environment and with the expectation that the human driver performs all remaining aspects of the dynamic driving task.

Level 2: Partial Automation. The driving mode-specific execution by one or more driver assistance systems of both steering and acceleration/deceleration using information about the driving environment and with the expectation that the human driver performs all remaining aspects of the dynamic driving task.

Level 3: Conditional Automation. The driving mode-specific execution by one or more driver assistance systems of both steering and acceleration/deceleration with the expectation that the human driver will respond appropriately to a request to intervene.

Level 4: High Automation. The driving mode-specific performance by an automated driving system of all aspects of the dynamic driving task even if a human driver does not respond appropriately to a request to intervene the car can pull over safely by a guiding system.

Level 5: Full Automation. The driving mode-specific performance by an automated driving system of all aspects of the dynamic driving task under all roadway and environmental conditions that can be managed by a human driver.

Levels 0–3 assume human presence in the vehicle and so parking will still require the same spacing as at present. However, with Levels 4 and 5, cars can park "on instruction" without a driver present. This means there would be no need to open the car door — meaning that bay widths could be reduced dramatically to just the width of the car plus an appropriate tolerance. However, if cars are parked closer together in this way, it could potentially increase the deck loading. In addition, there are concerns that electric vehicles present a greater fire risk than internal combustion engine vehicles, so it could be desirable to space vehicles further apart, rather than closer, to mitigate this risk (Section 10.3.1).

10.3 Car park design

10.3.1 Multi-functional use

The current trend is for car parks to become more akin to transport hubs where they include facilities for cycles, motorcycles and scooters — together with lockers for storage of equipment and clothing and amenities for showering and changing. Cycles and motorcycles do not mix well with other vehicles owing to the significant difference in sizes and velocities. Ramps also present an increased safety risk to those on two-wheeled transport and therefore they should be contained on the entry/exit area without the need to circulate the car park.

As electric vehicles (including motorcycles, cycles and electric scooters) gain popularity, the car park is the ideal venue for charging them — which requires careful consideration in respect of power supplies, ventilation and firefighting.

Autonomous vehicles are likely to have the greatest impact on car park design as they will be able to enter and self-park without the need for a driver and the occupants. Potentially this means that parking spaces can become smaller and storey heights can be reduced. However, depending on the fire strategy, bays may need to be spaced further apart for electric vehicles. Also, the requirements for internal finishes, stair cores and lifts will be less onerous, standards of lighting and ventilation can also be reduced, but not less than the requirement to ventilate areas where electric vehicles are being recharged.

Depending on the extent to which urban areas change, and how popular autonomous vehicles become, it is possible the car parks situated on the periphery will become popular as they will facilitate the need for fewer cars to enter the town or city centre. This will encourage use of alternative forms of transport from the car park thus improving human health and the local environment.

10.3.2 Entry/exit controls and payment systems

At present, most parking structures rely on some form of barrier control at entry and exit to manage access to the car park where the vehicle/driver will interact with the system to enter or leave. This could include the use of a remotely-read tag or by ANPR. Usually, if there is a payment process involved this is located separately from the entry and exit to minimise delays.

If a car park is accessible to the general public (and payment collection is necessary), or is private with access limited to authorised users, it is probable that a barrier-based system will continue to be used in most instances, simply because the logistics of protecting revenue and/or preserving site security militate against open access, free-flow alternatives.

What is likely to change in the future is the payment system. Future cars may have some form of electronic purse associated with the vehicle that will allow automatic payment of parking charges, tolls and even fuelling charges. Such systems are already in limited use in other countries. For example, Portugal's Via Verde toll system tag can already be read in many of the country's car parks and also used to pay car parking charges[71].

A further development of this philosophy will be needed to accommodate the anticipated availability of autonomous vehicles, where the driver would send the vehicle to park but not be present in person. The vehicle would have to be able to self-identify at entry and exit — both to demonstrate that it has access rights and to properly account for any fee due.

10.3.3 Speed control and containment

Increasingly, cars are fitted with driver-selectable controls, in the form of cruise control — which limits the speed of the vehicle without driver input from the accelerator — and radar systems that detect potential collisions and bring the car to a stop. These systems tend not to work at the speed levels experienced in car parks.

In 2022 EU Regulation 2019/2144 (the EU General Safety Regulation)[72] came into force. The regulation mandates that every new car must have speed limiters and dataloggers. It also requires all new cars to be fitted with an Intelligent Speed Assist (ISA) by July 2024. Although the UK is no longer in the EU, it will still comply with the new regulation. It is perhaps a not too illogical next step to see the development of external systems that would adjust the speed parameters according to local circumstances. Thus, in addition to restricting vehicles to the appropriate speed limit on the highway, it may be possible to install local controllers to keep speed inside the car park to a suitable level. If the car speed were to be managed in this way the potential for, and the seriousness of, any accidents would be reduced.

Containment is based upon errant vehicles striking a vehicle restraint barrier at a maximum speed of 10mph and the vehicle being brought to a complete stop by stalling the internal combustion engine. However, battery electric vehicles are heavier and less easy to stall and will continue to exert force on the barrier even though the vehicle is stationary, which is a consideration when designing the means of containment.

10.4 Services

Consideration for the futureproofing of building services provisions requires an acknowledgement that the typical design life of key building services elements will be significantly shorter than the structural elements of the car park in which they are installed. *CIBSE Guide M*[73] provides a guide to the economic life of building services elements and systems. In relation to car park design, key systems to consider are:

- Lighting (15–20 years)
- Power distribution (20–30 years)
- Ventilation (20–30 years)

Typically, these systems will have components with shorter life spans, but it is reasonable to expect these to be replaced without the need for significant alteration of the system in which it operates.

As a general guide, it is good practice to ensure that systems include some future capacity for unforeseen changes. This is to allow a degree of flexibility to adapt to small or moderate changes without complete system replacement. As a guide, 20% is a reasonable capacity to build into systems, although care should be taken so that this is not unnecessarily compounded across systems (e.g., 20% future capacity might be allowed on a ventilation fan/system but is not additionally required on the electrical distribution capacity to that ventilation system).

Certain future conditions may be considered as part of the building design approach. An example of this would be if a modular approach was taken, to allow for future physical growth of a car park (increased floor area, additional floors etc.). In this case, appropriate strategies should also be considered as part of the services design that will allow this to happen with minimal impact at a later date.

Another common example is making provision in the design for additional EV charging points. It is recommended that this is agreed with the client as part of the design strategy with consideration of current local authority or planning policies (Chapter 13).

Significant changes, such as a shift to fully autonomous vehicles, are likely to require reworking of key building services systems to accommodate, and it is likely to be uneconomical to attempt to futureproof the services for such events. Each building should be considered for its specific requirements, but key to any futureproofing is to consider all aspects – i.e., structural, architectural and building services to ensure a coordinated approach to any future provisions.

10.5 Fire considerations

At the very least, fire safety measures should meet the minimum fire safety standards that legislation and the fire safety strategy require, plus measures to counter any known future risk. Chapter 7 discusses some of the less well-defined fire risks from materials used in modern vehicles and the different risks between internal combustion enginned and electric vehicles. The fire design should be robust enough to allow for changes in the assumed risks which could reasonably be foreseen. Allowance should also be made for any known future changes in use.

10.6 Purposing and repurposing

'Purposing' is to design a car park solely for the purpose intended and in accordance with the principles of lean design (Chapter 11) to minimise the impact on the environment, the world's natural resources, and the amount of carbon generated. 'Repurposing' is a future change of use that is envisaged and either incorporated at the design stage or undertaken at some point in the future. The potential reuses are considerable and may comprise residential, office or leisure use, urban farming or hydroponics, power or self-storage etc.

Design for minimum carbon emissions can be at variance with discussions around future flexibility. With an expected service life of 50 years, a car park should be designed to minimise carbon emissions both today and in the future.

As a starting point it is preferable for a circular economy, to prepare a design that can be adapted/strengthened later, unless a distinct need to increase the flexibility today can be demonstrated[12]. Any increase in carbon emissions today must be determined and evaluated against the probability of a future change in use and the resulting environmental impact.

Unless a car park is only expected to be in service for a short time e.g., five years or less, it is unlikely that the nature of any future change of use can be predicted with any degree of certainty. Generalised assumptions around future flexibility, end of life disassembly scenarios, and future load increases should be avoided. Instead, the focus at each stage of design should be to assess how future use changes or load increases could be enabled through designing for adaptability, and how details and records could enable easy disassembly and reuse as part of a circular economy.

Car parks should never be overdesigned in a way that adds extra capacity for hypothetical future flexibility unless these are extremely likely scenarios that will occur in the near future and where the initial functionality of the car park is not adversely affected. If the car park were designed for maximum future flexibility, it would be necessary to consider the structural loading of the future use, the natural vibration required for the future use (e.g., most uses are more sensitive than car parking) and the structural fire resistance, among other factors. These factors could result in a far higher embodied carbon in the structure, which, if the building is ultimately never repurposed, is unnecessary and inefficient.

Instead, a strategic approach should be taken to enable the car park to be adapted and modified in the future to maximise future value, use and longevity e.g., allowing for localised strengthening, additional dampening, or localised disassembly and relocation (Chapter 8). This approach is typically more materially efficient, as well as mitigating the climate impact of accumulative greenhouse gas emissions — by delaying intervention to a time when materials are available, with a lower environmental impact.

Part 2
Detailed design

11 Design and construction

11.1 Loading

11.1.1 Introduction

In Europe, the imposed loading applicable to decks and ramps in car parks is given in BS EN 1991-1[74] with the appropriate National Annex in the UK[75]. While these loadings are based on a maximum vehicle weight of 2,500kg and an average weight of 1,500kg, weights have tended to increase with the introduction of hybrid and battery electric vehicles, the increase in size of vehicles and the greater number of SUVs. Currently the average weight of 95% of vehicles on UK roads is 2,000kg and the maximum weight of large cars and SUVs can be more than 3,000kg.

11.1.2 Uniformly distributed loads

The uniformly distributed imposed load given in BS EN 1991-1 — for a standard mix of vehicles, subject to the maximum gross weight (weight of vehicle plus passengers and luggage) of any vehicle up to 3,000kg — is 2.5kN/m^2. This figure includes an allowance for vehicles in motion imposing a greater load on the structure than when stationary due to vertical oscillation on their suspension.

In the knowledge that average and maximum weights of vehicles are increasing, it is recommended for good practice that the uniformly distributed imposed load is increased to 3.0kN/m^2.

In certain situations, vehicles can be parked in smaller bays e.g., by automated robotic systems (Section 20.2) or valet/block parking (Section 20.3) and it is recommended that the uniformly distributed load is increased in proportion to the density of parking.

For multiple floor loads it is recommended that the approach in BS EN 1991-1-1 is adopted, which for car parks does not allow any reduction in column load from loaded areas and requires that all storeys are assumed to be simultaneously loaded.

For other countries there may be local loading variations which should be verified by reference to local codes/standards.

Local authorities often require fire engine access to designated areas, particularly at roof level. This requirement should be established at an early stage since it has a significant effect on the roof structure, the ramps, any intermediate floor access zones, and columns. Note, that while the fire engine access routes will be agreed in the event of an incident, the fire tender (apparatus used to transport water from source to scene of fire) may inadvertently take the most direct route.

Snow loading on roofs is not usually considered in combination with vehicle loading, unless the car park is being designed for long-term parking or is in an area of high snowfall. Consideration should be given to the type of snow-clearing equipment used, as some equipment can be heavy and overload the roof deck and ramps.

Cladding loads may be influenced by icing, particularly on exposed grillages.

11.1.3 Wind loads

Wind speeds and loading are determined from BS EN 1991-1-4[76] and other literature[77].

Percolation through the building will result in wind drag on the bodies of the parked vehicles and will be transferred to the floors by shear in the tyres. It is therefore recommended that the design wind loading is taken as acting over the entire elevation area of the structure with no reduction for openings. Wind drag must be considered in structures that are susceptible, particularly those with lightweight steel floors, and for temporary and demountable car parks (Section 11.11).

Local wind pressures on cladding components must be designed on the basis that maximum pressure and suction is applied simultaneously to both faces of each component.

11.1.4 Lateral loads from vehicles

In addition to wind load, car park structures should be designed to withstand vehicle impact loads. Impact may be from within the car park (either directly or by transmission from the edge protection barriers (Chapter 12)) or from outside the car park at ground level. Section 4.3 of BS EN 1991-1-7[78] provides equivalent static design loads for a wide range of vehicular impacts against structures.

Lateral loads also arise from vehicles changing direction vertically at ramps and from friction when they turn or brake. Although rarely critical in permanent structures they need to be considered for temporary and demountable structures.

Consideration of lateral loading is more important in car parks as they do not derive additional stiffening from partitions and finishes. Lateral loads should be safely transferred to the foundation through the structural system e.g., slabs, beams, bracing and columns.

Lift systems and escalators/travelators apply significant additional loads to localised areas of the structure, so provision for their inclusion should be established at an early stage.

Allowance must be made for access and any abnormal loads for maintenance, firefighting, and snow clearance.

11.1.5 Ground pressures

Flotation may dominate the design and construction of basement and semi-basement parking levels. The most critical flotation case is often during the early stages of construction. Provisions for drainage of the basement and of the ground below and around the basement must be considered.

Lateral soil pressures on the walls, uplift and ground heave pressures under the groundbearing slab will dominate the section geometry and design of the basement and semi-basement parking levels. The magnitude of such pressures is likely to be several times greater than common permanent or variable actions on the floor slab.

11.1.6 Punching shear from jack or wheel loads

Decks less than 150mm thick should be designed to carry the more onerous of the following:

- Wheel loads as shown in Table 11.1. The figures are based on the maximum tyre pressures recommended for given tyre sizes. However, BS EN 1991-1 recommends an axle load of 20kN spread on two 100mm squares at 1.8m centres
- For vehicles up to 3,000kg a jack load of 15.0kN acting over 50mm × 50mm. This contact area need not be considered closer than 0.75m to a floor edge

If traffic-calming measures such as speed humps are used, it must be demonstrated that the increased loading due to impact does not exceed the normal design imposed loading. It is generally preferable to use signage or tactile paving to avoid impact loads on slabs, maintain headroom, and to avoid discomfort to vehicle occupants.

Table 11.1: Wheel loads

Contact area (mm)	Imposed load on flat slab (kN)
150 × 150	6.5
175 × 175	9.9
200 × 200	15.0

11.1.7 Load combinations for common design situations

Loading on car parks is generally similar to that on other buildings. However, the ease with which imposed load can be moved makes the analysis of pattern loading cases important. Thermal strains may also induce forces that can be significant if restrained, particularly for exposed roof slabs and parts of the structure that are incorporated to provide stability i.e., bracing, lift and stair cores etc.

Structural design and detailing should be carried out with critical combinations of permanent actions, variable actions, snow load, wind load, temperature effects (Section 11.3.4), the notional variable actions from relevant parts of BS EN 1991[74, 76] and the other lateral loads.

Design against flexure, shear and torsion should be carried out, using design methods and partial safety factors according to the appropriate codes of practice for the structural material and construction type.

11.1.8 Load combinations for uncommon design situations

The loads given in Sections 11.1.9 and 11.1.10 should be considered using the appropriate combination of partial factors. Note, that where access to the car park for emergency vehicles such as fire engines or maintenance vehicles is required, the design loading may have to be increased.

11.1.9 Resistance to explosions

Explosions within car parks may be caused by liquid or gaseous fuel, or from chemical explosives. Because car parks have open plans and considerable venting to extract fumes, the effects of liquid fuel or gas explosions are unlikely to be critical to the overall structure, though they may make individual structural components ineffective. In the rare event that a car park must be designed to resist chemical explosions, appropriate loads are given in BS EN 1991-1-7.

Where gas is supplied within the car park, the structure should be designed to cater for a potential explosion and to prevent disproportionate collapse.

A designer's risk assessment should be considered if the car park is beneath occupied floors of a building or it exceeds 6-storeys, as fire is more likely to be critical[79].

11.1.10 Landscape loading

A wide variety of loading may result from different landscaping schemes for roofing and walled areas, and this must be carefully considered. In deriving the appropriate loading, account must be taken of the planting scheme required, mature size of plants, maintenance, and the flexibility for future changes.

The imposed loading from landscaped areas, water features and concentrated tree loads can be an order of magnitude higher than the normal vehicular imposed loading for car parks. Even where thin, drained grassed areas are proposed, it is recommended that a minimum imposed load of 7.5kN/m^2 be adopted. Similarly, loads from water features and rock features must be carefully considered as they can add significant loading, limit flexibility for future changes, increase structural costs, and require a continuing operation and maintenance commitment.

The provision of tree planters (Figure 11.1) in particular must be carefully planned, as the increased depth of soil required (typically at least 1.5m), the weight of the mature tree, and overturning effects of wind loads all lead to concentrated and local loading. In the absence of more specific details, it is recommended that a percolation factor of 50% be used in conjunction with the projected elevation of the fully mature tree to derive wind forces, and that they are placed over column and/or principal support positions.

The features of landscaped areas will usually include provision for drainage, irrigation and access for maintenance. Owing to the difficulties of access to inspect membranes and the need to both irrigate and drain landscaped areas, such areas will usually require an enhanced waterproofing membrane and a root barrier, or positive measures to prevent structural damage by root penetration.

The membrane must be adequately protected to prevent damage from horticultural maintenance. The soil must be properly drained, without being too steep, and falls of 1:100 to 1:40 would normally be used in conjunction with a

Figure 11.1: Minimal roof planting details

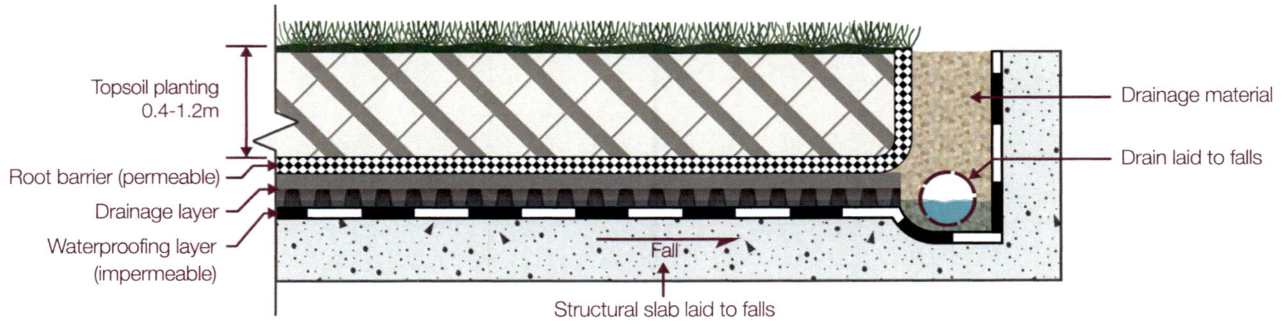

positive drainage system to ensure the soil does not become saturated (Fig. 11.1). Loading must consider the effects of blockages to the drainage system and the increased loads resulting from storm inundation.

Provision of edge barriers, means of access, and means of avoiding steep slopes requiring frequent maintenance must be properly dealt with so that landscaped areas can be safely maintained.

The restrictions on loading, the requirements to control growth, and the maintenance regime to ensure loading does not exceed the design level, must be clearly defined and agreed with the client as part of the brief. This information is an essential part of the operations and maintenance manual, the life-care plan and the health and safety file.

11.2 Other considerations

11.2.1 Disproportionate collapse

Car park structures should have provisions for robustness as given in the relevant codes of practice, design standards and related guidance[6] for measures against collapse disproportionate to the cause, e.g., a local accident, effects of fire, etc.

In the UK, Building Regulations Approved Document A3[5] categorises the consequence class of car parks and the measures required to avoid disproportionate collapse. Car parks not exceeding 6-storeys are Consequence Class 2b, those greater than 6-storeys are Consequence Class 3.

Section 3 of BS EN 1991-1-7 provides recommendations on how to identify potential accidental actions, and how to limit the extent of consequent local failure.

11.2.2 Response to vibration

Human perception of vibration in buildings varies. It also varies depending on the activity being undertaken, as stationary people are more aware of vibrations than those who are moving. Likewise, those engaged in sensitive work require an environment that is not susceptible to vibration.

Empty car park structures lack the damping normally provided in buildings by partitions and finishes. However, the dynamics of most car park structures are generally found to be satisfactory when the design gives natural frequencies above 5Hz, even when compartments within or adjoining the car park are in use as offices. Where car parks form part of a mixed-use development, measures to reduce vibration of the interface or shared floors should be enhanced to suit the more onerous condition as defined in BS 6472[80].

User perception of vibration is lessened by the user's motion either when walking or travelling in a vehicle. In standalone car parks with lightweight construction and spans up to 16m it has been found that a minimum natural frequency of 3Hz is satisfactory. This is calculated for simply supported primary beams and floor slabs, together with secondary beams where appropriate within the effective width.

Cars parked on the decks provide additional damping to the structure by their mass being supported on damped suspension. Any end restraint will assist in damping the vibrations and its contribution should be considered if justification of the vibration characteristics is by *in situ* testing of the completed structure.

Unusual or long-span structures may be more sensitive to dynamic loading, and their response to vibration should be checked. Irregular surfacing, badly aligned movement joints and speed humps can have an adverse effect on dynamic movement.

11.3 Designing for movement

11.3.1 General

Structural elements of car parks are susceptible to movement both during and after construction. Some of the principal reasons are:

- Elastic structural deflections
- Accidental loading
- Temperature changes during construction and in service

- Shrinkage
- Creep
- Differential settlements

With careful design, appropriate joints can be provided to accommodate such movement and to suit the structural design. Car parks can suffer from premature deterioration if the sealants break down under traffic loading, allowing water and salts to enter the joints (Section 11.9).

Allowance should be made for the fact that the exposed nature of car park structures produces a greater design temperature range than in other building structures. Joints must also take account of the potential for differential movements between the relatively flexible car park deck and the stiff zones created by ramps and access shafts.

Typical mastic joint fillers can only accommodate around 20% strain, so a mastic-filled joint would typically need to be 5x wider than the predicted movement. Elastomeric joint fillers are generally regarded as more effective for joints with large predicted movements. The modulus of the joint fillers should be selected to account for the cyclic nature of the predicted movement, such as the diurnal and annual temperature variations. Their additional cost must be balanced against potentially expensive remedial work in the event of joint sealant failure.

Several reports published by The Concrete Society[81–83] provide valuable guidance on various causes of cracking, which can lead to premature deterioration of structural elements. Structural design should also aim to control such cracking, including that resulting from differential strains due to restrained movements at the interface between pours or between precast and *in situ* concrete, for example.

11.3.2 Deflection

The prediction of long-term service deflections is of great value — to confirm that falls are and will remain sufficient to prevent the ponding of water on surfaces or in drainage channels.

In normal circumstances, falls of 1:60 are considered adequate to accommodate construction tolerances, pre-camber and the effects of deflection under short and long-term load. For long-span beams or transfer structures, detailed calculations of deflection should be undertaken to ensure the slab drainage strategy is not compromised.

11.3.3 Temperature

Multi-storey car parks differ from other buildings in that they have a structural frame that is usually exposed. This means that the effects of temperature-induced movement must be specifically considered and catered for.

Car parks can be considered as intermediate between bridges (where temperature loads are well codified and established) and standard building frames where the structural frame is normally within a controlled environment.

Temperature movements are restrained by the lateral stability systems. Careful consideration should be given to the location and interaction of such systems to avoid large forces being generated by restrained thermal movement. In concrete structures, these can lead to load reversal, causing cracking of elements not designed to resist the stresses induced.

Constrained expansion often occurs inadvertently between the deck and ramps or stair structures. If the adverse interaction between lateral restraint systems cannot be avoided, then either the system should be isolated from the slab, or a movement joint should be introduced to limit the movement. Care must be taken to provide for any lateral load transfer function to be maintained.

Temperature differential through exposed top decks will cause thermal bowing in summer. This requires consideration when deck joints are being configured.

Stair cores and surrounding buildings can cause shading on exposed top decks as a result of the sun's movements. This can lead to cyclic temperature variations which should be accounted for in the design.

11.3.4 Temperature effects

The serviceability implications of temperature-induced movement and restrained cracking in structures require consideration, as excess joint movement causing leakage and cracking can lead to durability problems. The design temperature range that may need to be considered can be determined from BS EN 1991-1-5[84] and the IStructE's *Manual for the design of building structures to Eurocode 1 & Basis for Structural Design*[85].

The moments induced in columns by heating and cooling of the top slab relative to slabs below, as well as changes in reactions from the differential within the top slab, can be significant in flat slab punching shear.

Thermal bow for simply supported units should be considered. For an initial estimate it can be taken conservatively as:

$$\frac{(\alpha \times \Delta T \times L^2)}{8 \times t}$$

Where:

α = coefficient of thermal expansion

ΔT = temperature difference

L = length

t = thickness

More detailed guidance can be obtained from the Eurocodes e.g., BS EN 1992 for concrete structures[86].

In general, steel and concrete frames have a similar coefficient of expansion. This may be taken as $12 \times 10^{-6}/°C$.

Certain types of aggregate, e.g., limestone and granite, have lower coefficients of expansion[87] and could be considered for use as part of the design.

Temperature effects may be introduced into the design[74, 85].

Movement joints must make suitable allowance for the effect of temperature on the joint width during construction. Some mastic joint fillers may not be suitable for multi-storey car park construction, since their movement accommodation factor is limited.

In order that all these issues are considered in a rational way, the following procedure is appropriate:

1. Establish proposed basic car park form and elevational treatment, lateral stability system, ramp and stair locations.
2. Calculate design temperature of top deck frame.
3. Establish likely mean temperature during construction, and hence the temperature range.
4. Taking account of both the lateral stability system and any other secondary lateral restraint present, calculate frame movements at elevations and other critical features.
5. Check that any constrained expansion forces are acceptable.
6. Check that frame movements are compatible with elevational treatment.
7. If steps 5 or 6 are unsatisfactory, either revise layout or introduce movement joints.
8. If movement joints are adopted, recalculate frame movements.
9. Re-check that frame movements are compatible with elevational treatment and proposed movement joint details.
10. Check top deck thermal bowing effects.

11.3.5 Shrinkage

Shrinkage in concrete must be accounted for in the design of the structure[86] considering specifically restraint of relatively thin decks by solid cores that can result in cracking. The characteristics of shrinkage and creep vary in different countries and climates.

In general, precast concrete deck and frame elements will have largely completed their shrinkage cycles by the time they are incorporated into the structure. However, shrinkage must be considered in the configuration and sequencing of *in situ* concrete frames and decks, or any system with a continuous *in situ* concrete wearing screed. Post-tensioning, if used, will cause elastic shortening; the effect of which must be taken into account.

Shrinkage in prestressed concrete units can generate high bearing stresses at the supports, and when used in conjunction with a concrete frame can result in spalling of the support — especially when there is no bearing material present, and the bearing area is small owing to the precamber generated from prestressing the units.

Frame shrinkage must be considered and accommodated by joints in brittle-facing materials such as brickwork. Joints in facades are normally required at much closer centres than those in the principal structure. For example, brickwork may require joints at 9–12m centres, and horizontal supports may also be required to restrain brickwork panels.

11.3.6 Creep

Creep can have a significant effect on the deflection of reinforced or prestressed concrete units over time. Methods for calculating its effects are well documented in BS EN 1992-1-1. Creep can be more significant for concrete car park construction than other structures for the following reasons:

- Column spacing is usually maximised to improve vehicle circulation. This often leads to long-span prestressed floor designs
- The ratio of permanent actions to total load tends to be higher in car parks than in other structures. A typical concrete car park when empty can be 75% of its gross characteristic design weight. This compares with approx. 50% for typical office structures. A high proportion of the variable load is also achieved in service

Creep significantly modifies the effective elastic modulus of concrete beam members over time. This can mean that members whose deflection performance is adequate shortly after construction can be inadequate after perhaps 10–20 years. Deflection of concrete long-span beams can increase by approx. 25–50% which can adversely affect drainage provisions. It is therefore essential that their effect on the drainage strategy at the design stage is fully considered. It is usual to locate gully outlets and downpipes at column positions. Creep and consequential long-term deflection of concrete beam members can lead to ponding at beam mid-span or cantilever tip locations.

11.3.7 Movement joints

Adequate allowance must be made for car park structures to respond to temperature-induced volumetric change and for shrinkage of reinforced or prestressed concrete structures[83]. Typical movement joints for use in external decks are shown in Figure 11.2. External decks are subject to greater movement, and in this situation a proprietary waterproof movement joint would be more suitable.

Figure 11.2: Typical movement joints

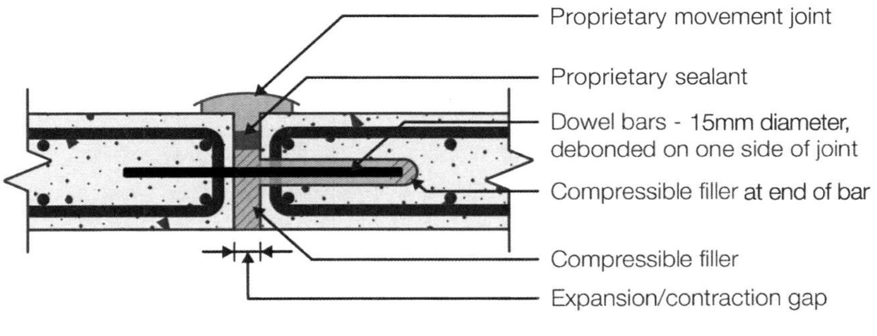

a) Dowelled movement joint

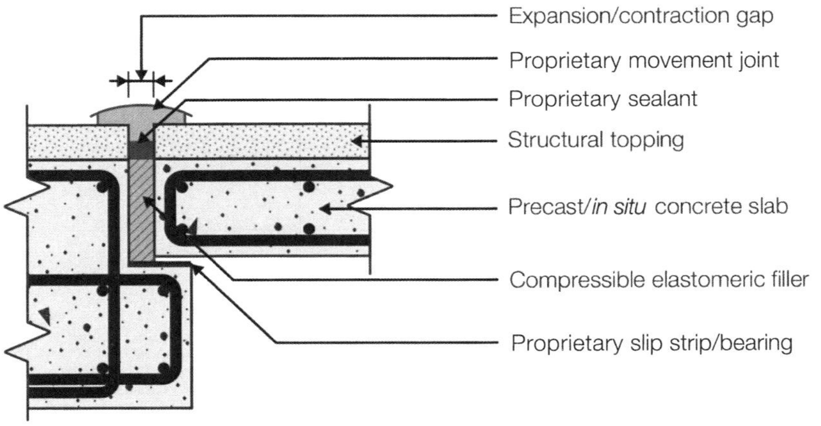

b) Sliding movement joint

Installation of movement joints must also take account of the season of construction. Ideally, joint widths should be adjusted during construction to suit the prevailing temperature conditions and seek to minimise the range of post-construction movement. As with all joints, provision should be made for easy replacement of components, as the design service life will be less than that of the main structure. BS 6093[88] gives guidance on the design of joints in building structures.

11.4 Structural materials

11.4.1 Introduction

Concrete and steel — sometimes combined in a variety of ways — are conventional materials used in car park construction. Concrete may be both reinforced and prestressed. Steel is used either alone as the principal structural material, or compositely with concrete.

11.4.2 Concrete construction

Concrete car parks may be assembled from precast units or cast *in situ*. *In situ* concrete structures may be cast on a wide variety of proprietary formwork and falsework systems; proprietary lift-slab methods avoid the need for formwork for some types of car park.

The use of lightweight-aggregate concrete can also be considered. In car parks, the weight of the concrete slab usually exceeds the variable action, and a 25% reduction of this weight is significant, both in the slab and in its effect on column and foundation loads. Lightweight concrete also offers better fire performance. This must be set against the disadvantages: a smaller permitted span:depth ratio, additional shear reinforcement in the slab, and in composite beams a slightly reduced effective breadth of flange and an increased number of shear connectors. Lightweight concrete generally provides less durability, and implications of its use on long-term performance and its response to vibration should be considered.

11.4.3 Steel construction

Unprotected structural steelwork (i.e., without a fire-protective coating) is often used in car park structures. Its use in above-ground car parks can sometimes be justified using fire engineering analysis techniques, based on ventilation conditions and perhaps utilisation ratios (Chapter 14).

When fire protection is provided to steelwork, the appropriate atmospheric corrosivity category must be specified as the frame is exposed and systems must have adequate strength to resist minor impacts (e.g., scrapes). Systems whose performance in fire is degraded by scrapes should be avoided.

11.4.4 Composite construction

Car parks in composite construction generally comprise a framework of steel beams and columns supporting a concrete floor (Figure 11.3). The floor usually combines in composite structural action with the steel beams in one or both directions and precast units with grouted joints and an *in situ* wearing screed (yet to be applied in Fig. 11.3). Note, for industrial uses the wearing screed can be omitted. *In situ* concrete can also be used, cast on removable formwork or on metal decking that serves as a permanent shutter. Some of the advantage in speed of erection afforded by prefabrication may be lost if wholly *in situ* construction is adopted for floor slabs.

11.5 Designing for construction

11.5.1 Above-ground car parks

11.5.1.1 Introduction

The structural form of a multi-storey car park will be heavily influenced by the design geometry adopted. Designs commonly use *in situ* concrete, precast concrete, structural steel, or a combination of these. Guidance on the location of columns is given in Section 5.3.7.

Many construction sites have restricted access and spacial constraints, such as existing buildings and road and rail infrastructure. The choice of structural form and materials should recognise and address all logistical constraints.

Figure 11.3: Composite construction

11.5.1.2 Floors

11.5.1.2.1 General

With the exception of some temporary structures, car parks use concrete for the deck structure. It is recommended that all decks are laid to a minimum 1:60 fall to facilitate drainage and minimise the ponding of water.

Exposed decks usually have a waterproofing layer. This can be omitted through the use of watertight concrete, or deck construction methods that conform to water-retaining design standards provided there is adequate frost resistance. Water retaining design standards do not guarantee zero water penetration but will generally reduce crack widths. As with any system, care must still be taken with the design and construction of movement joints.

High chloride concentrations can occur on car park decks, owing to deposition of vehicle-borne salt combined with poor drainage, detailing and maintenance, including use of de-icing agents to prevent slip incidents. This could cause premature reinforcement corrosion, particularly in top steel for slabs continuous over supports and especially for exposed decks. Areas close to vehicle entry points also have increased vulnerability. Refer to Section 11.10 for more information on durability issues.

The car park decks must also be designed to resist horizontal and vertical loads applied by cantilevered vehicle edge-protection systems and their fixing arrangements.

The main types of floor construction are:

- Precast concrete hollowcore slabs with *in situ* concrete wearing screed
- Precast concrete permanent formwork with *in situ* concrete wearing screed
- Precast solid slab
- Precast concrete double tees
- *In situ* concrete flat slab
- *In situ* concrete ribbed slab
- Metal decking permanent formwork with *in situ* concrete slab
- Steel decking

Each is discussed in detail in the following subsections.

11.5.1.2.2 Precast concrete hollowcore slabs with *in situ* concrete wearing screed

These are mass-produced pre-tensioned voided units produced by long-line methods and sawn to length on the casting bed. They use high-strength concrete and usually have high standards of quality control. Units with span capability up to 16m are available (Figure 11.4).

Figure 11.4: Hollowcore slab details

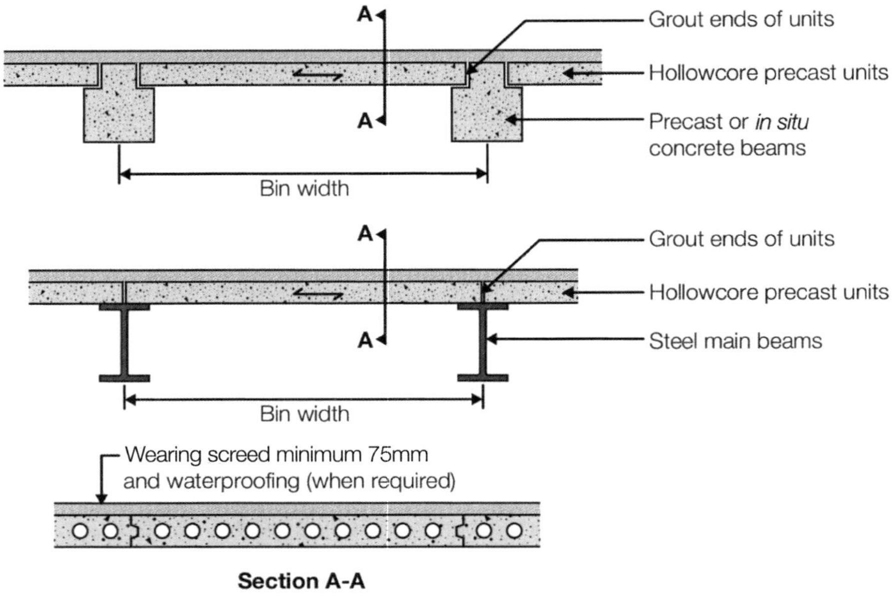

If hollowcore units are used, the following issues should be given careful consideration:

- They have neither lateral distribution reinforcement nor lateral strength — and diaphragm action is usually provided by grouting the sides together to form a shear key together with a bonded reinforced *in situ* concrete wearing screed. The screed also provides a suitable surface for application for waterproofing and running of vehicles
- Slab bearings require careful consideration. Hollowcore units on uneven bearings have been known to split longitudinally to relieve transverse tension across the units induced by poorly aligned seatings
- Excessive deflection of the supporting beams can also lead to longitudinal splitting of the hollowcore units
- Concrete ledge seating should be configured to ensure the loading (including the load distribution) is confined within the reinforcement cage to avoid spalling when the hollowcore units deflect cyclically under variable or thermal actions (Fig. 11.4)
- There is no concrete cover to the ends of the prestressing strands in hollowcore units. Suitable end protection must be specified to prevent corrosion
- Hollowcore units have hidden voids and are often plugged at exposed ends to exclude water or to enhance their shear capacity. The voids can still fill with water, either during erection or from failure of waterproofing membranes (Section 11.10.6). It is unlikely that prestressing strands will have adequate cover for durability around the cores, leading to possible deterioration
- Weep holes must be provided in hollowcore units to ensure water does not collect, as it can freeze and split the units, and increase the load. Manufacturers can provide weep holes to drain trapped water, or they can be drilled on site retrospectively. Wherever possible, weep holes should be positioned at the low points when the units are laid to falls
- Inspection of the cores is very difficult, if not impossible. Bridge engineering has moved away from permitting hidden voids because of the problems of corrosion due to saltwater ingress[89]
- Water may drip from the weep holes onto parked vehicles, leading to damage to paint and glass. Guttering may be required where leakage occurs
- The hollow form means the top surface can heat and cool more quickly than the underside when exposed to sun. This can cause bowing — damaging the bearing area, and splitting the membrane

11.5.1.2.3 Precast concrete permanent formwork with *in situ* concrete wearing screed

Typically solid and 50mm or 75mm thick, this type of floor construction may be reinforced with steel fibres or pre-tensioned (Figure 11.5). These units are designed to span between primary beams at up to 5m centres, but may require propping during construction, limiting applications.

11.5.1.2.4 Precast solid slab

This form of construction is used in proprietary systems to achieve some of the thinnest decks. Although the span of the slab is limited, it is an economic solution when designed to act compositely with long-span steel beams; achieving diaphragm action (Figure 11.6).

Figure 11.5: Precast concrete permanent formwork

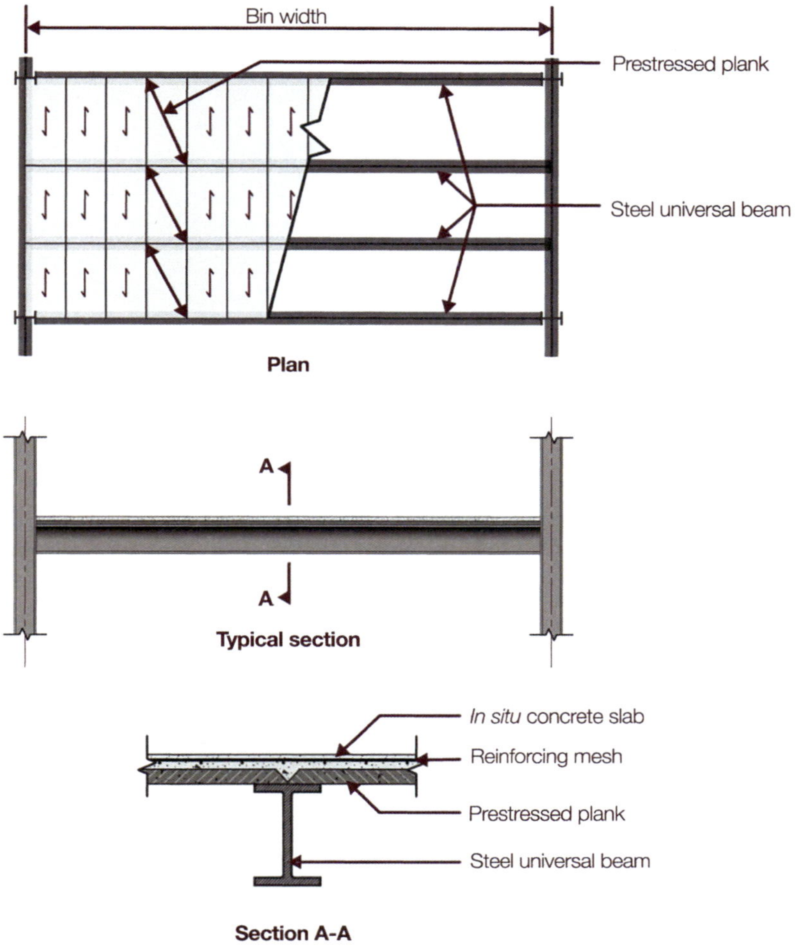

Figure 11.6: Precast solid slabs

11.5.1.2.5 Precast concrete double tees
Double tees (Figure 11.7) are cast using a long-line pre-tensioned method, but individually shuttered with stop ends to form the unit ends. They are typically 600mm deep for 16m clear span. They may be used with an *in situ* wearing screed (a top flange depth of 75mm is common) or without — in which case the top flange thickness is likely to be at least 100mm. Shear connectors are typically used to achieve diaphragm action. If there is no *in situ* wearing screed the durability of the connectors needs to be considered in the material specification.

The long-line process means that it is necessary to provide *in situ* protection to the exposed reinforcing strands at the unit ends. If constructed with no *in situ* wearing screed, great care has to be taken to accommodate differential pre-cambering and waterproofing.

Figure 11.7: Precast concrete double tee

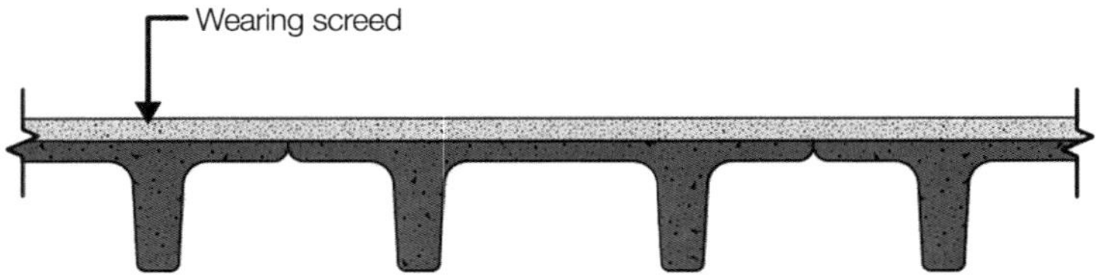

11.5.1.2.6 *In situ* concrete slab
Flat slab construction is often proposed for car parks as it can reduce clear floor heights — and for basement car parks it can be used to brace contiguous pile walls. The slabs can be reinforced or post-tensioned uniform thickness or a waffle construction with columns spacing up to 8m with two-way-spanning slabs to achieve an economical solution. The impact of these columns on the parking layout needs careful consideration. There are proprietary systems (e.g., lift-slab) in which slabs are stack-cast at ground level and jacked into position, as well as long-span post-tension solutions.

Careful detailing and construction are required at the slab-to-column interface to prevent cracking and/or moisture paths that can promote corrosion of critical joints, which can lead to premature deterioration and failure (Section 11.10).

11.5.1.2.7 *In situ* concrete ribbed slab
Ribbed concrete slabs spanning one way (Figure 11.8) can easily be configured to deal with both long-span and short-span solutions. Usually they are of reinforced construction, but post-tensioned solutions are also possible.
It is important that continuous designs pay proper attention to the possibility of chloride ingress, as the main slab reinforcement will be at the top of the slab at the supports.

11.5.1.2.8 Metal decking permanent formwork with *in situ* concrete slab
Galvanised steel profiled metal decking (Figure 11.9) can be used as permanent formwork but is most commonly designed to act compositely with *in situ* concrete to form the slab. While regularly used in offices and other dry environments, it is not ideally suited to the aggressive car park environment.

The variable thickness of the profile results in thin areas of concrete that are prone to crack. Measures must be taken to prevent water ingress through the concrete, both during and after construction. Failure to do this will provide a path for water and de-icing salt run-off to penetrate down to the upper surface of the metal decking. This can cause premature corrosion, and to reduce the risk the upper and underside of the metal decking should be treated with a specialist coating by the manufacturer, and the running surface of the concrete should be coated with a waterproof membrane. Exposed roof slabs are particularly susceptible to such effects.

If it is to be used, all cracks must be sealed and a waterproof membrane applied to the concrete. It must be regularly inspected for damage, with repairs undertaken promptly. For these reasons, metal decking permanent formwork should be used with caution (Section 11.10.3.2.1).

Condensation can occur on exposed steel surfaces when temperatures incrase rapidly.

11.5.1.2.9 Steel decking
Galvanised structural steel decking may be suitable for car park floors and has certainly been used for temporary car parks, which have a significantly reduced service life compared to permanent car parks. Its long-term durability needs

Figure 11.8: *In situ* concrete ribbed slab

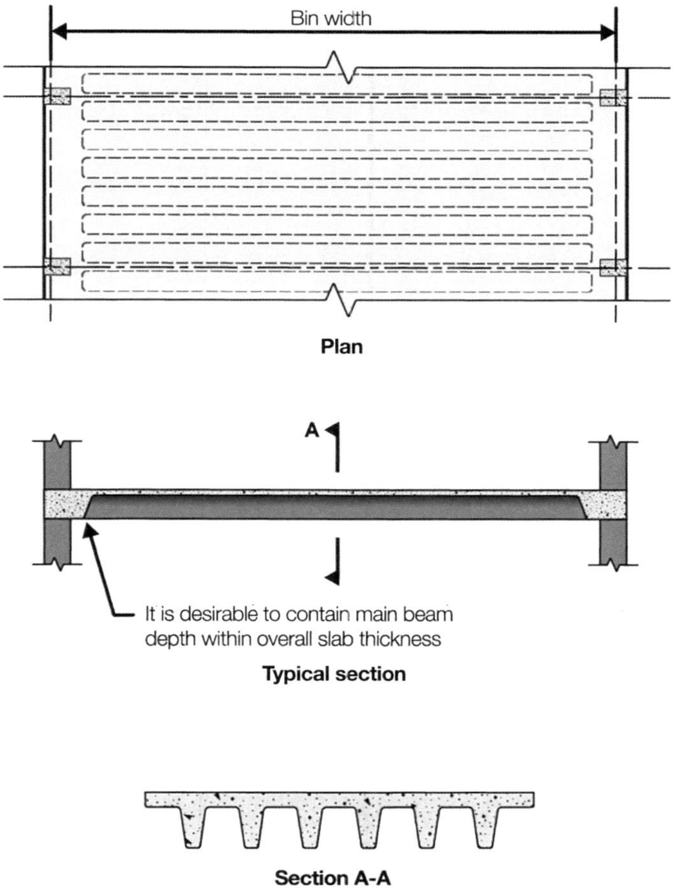

Figure 11.9: Metal decking permanent formwork

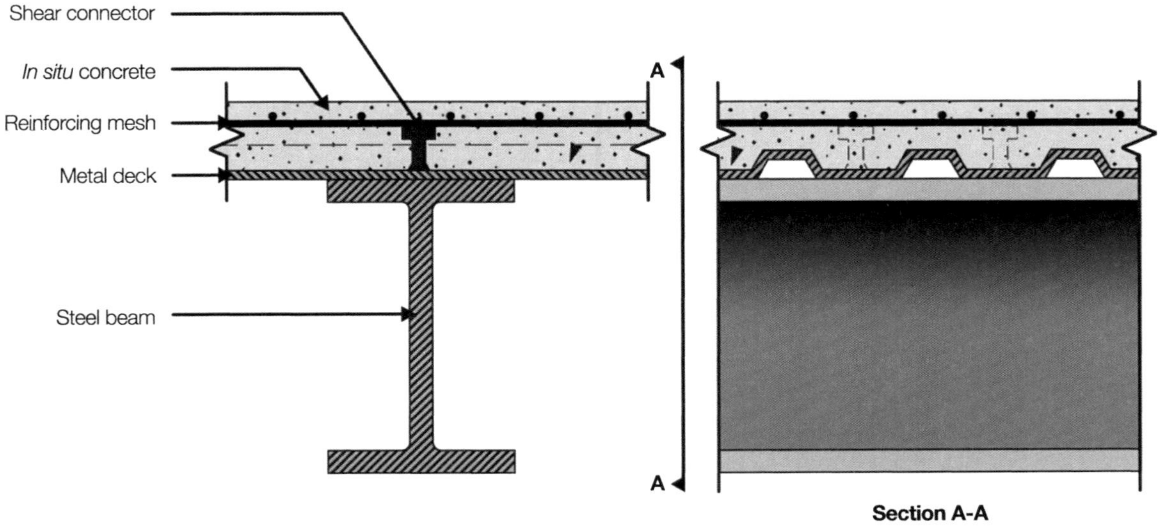

careful consideration if used in permanent car parks. Similar issues can exist with crevice corrosion on the top flanges (Section 11.5.1.2.8).

To ensure adequate skid-resistance for vehicles and pedestrians and to enhance its longevity, a waterproof membrane is usuallly applied to the steel decking.

Steel decks tend to be thin and can be sensitive to vibration, and they also tend to transmit noise while in use.

11.5.1.2.10 *In situ* concrete wearing screed
An *in situ* wearing screed is commonly used in conjunction with steel and precast concrete structural floor components (Figure 11.10). Steel reinforcement in the form of welded fabric and/or reinforcing bars should always be incorporated. It is important that details are developed that take into account such issues as:

- Reinforcement details at sheet overlaps to maintain specified cover
- Pour sizes and shrinkage
- Concrete grade and mix design (Section 11.10)
- Influence of camber and/or deflection on wearing screed thickness
- Asymmetric deflection of slab as wearing screed construction progresses
- Post-construction fatigue of wearing screed under load cycling where adjoining structural floor components have different flexural stiffness
- Workmanship during construction; in particular concrete compaction and support of reinforcement to prevent displacement during pouring and finishing operations. Structural wearing screed thinner than 75mm can be difficult to construct and if thinner than 60mm can result in curling at the edges

Figure 11.10: Composite beam: precast concrete slab and wearing screed

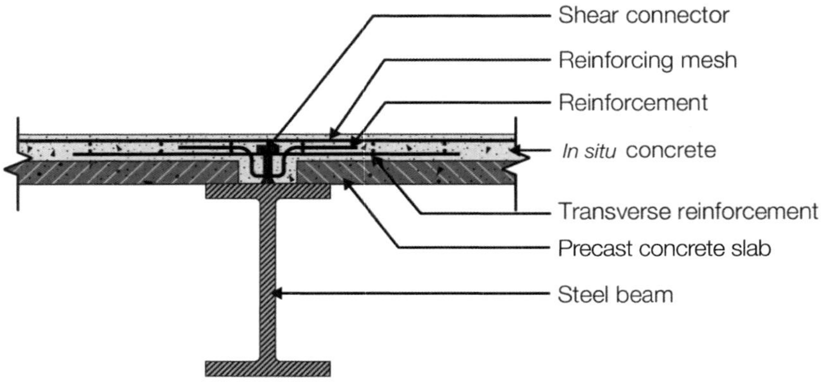

11.5.1.3 Frames

11.5.1.3.1 Introduction
The design solution for the floor slab and the constraints on the column positions (Chapter 5) will greatly influence the framing type adopted (Figures 11.11–11.13). Each principal type of floor construction usually has a compatible framing system, being either:

- Structural steel[29]
- Precast concrete[29, 90]
- *In situ* concrete[29, 90]

11.5.1.3.2 Integral framing solutions
The framing systems for an *in situ* concrete flooring solution are designed to be an integral part of the floor slab. This is also true for steel frames which incorporate steel beams and composite concrete slabs, including proprietary precast slabs that act in composite with the steel beams.

11.5.1.3.3 Separate framing solutions
Separate frames are typically used in conjunction with precast floor construction. The precast floor slabs are characterised by long-span decks with the steel or concrete support frame span minimised to accommodate the large

Figure 11.11: Steel clear span arrangement

Figure 11.12: Precast concrete clear span arrangement

Figure 11.13: Alternative cantilever frame arrangement

end reactions. A primary concrete beam can either be part of the precast system or *in situ* concrete. A deck span of three parking bays is typical for a precast slab, but a hollowcore unit can achieve spans of approx. 16m.

Cranked precast floor units can be effective in reducing secondary framing at ramps. With arrangements using long spans or split-level floors in combination with short columns, particular care is required to prevent local cracking of the column.

Care should be taken in the design and detailing of ends of slabs and secondary beams bearing directly onto primary beams without bearing materials. A sagging secondary member may bear on the edge of the primary beam, risking spalling.

11.5.1.4 Foundations

The design of foundations should follow standard practice[91, 92]. Guidance on this may be obtained from *The allowable settlements of buildings*[93] and *The real behaviour of structures — soil-structure interaction*[94]. Care must be taken when considering articulation of any potentially stiff continuous members to avoid cracking of concrete or unacceptable structural distortion in steel.

It should be borne in mind that loadbearing ground-floor columns and walls may not be restrained by the ground-floor structure and may be subject to internal and external vehicle impact. Foundations must be designed to have sufficient lateral resistance to these loads.

The lighter weight of a steel-framed structure compared with a concrete frame can have a significant effect on foundation costs in certain soil conditions[95, 96]. The reduced weight coupled with the structural flexibility provided by this form of construction can permit column loads to be carried by pad foundations. A heavier rigid building on the same site may require some form of ground improvement at pad locations or a different foundation solution, such as piles or a rigid raft.

11.5.1.5 Ground floors

Several options are available for ground-floor construction. The choice of a particular type will depend on groundbearing capacity, relative levels and client preference. The different types are summarised in the following sub-sections.

11.5.1.5.1 Asphalt

A flexible asphalt construction will usually be economic where low groundwater level and reasonable groundbearing capacity exist and where the client has no objection to a dark-coloured finish. In general, asphalt finishes must be laid after completion of the superstructure; limited headroom will restrict mechanised laying techniques.

Foundations levels should be set at a minimum of 300mm below finished levels to prevent the foundation outline reflecting in the asphalt. In locations where the above-ground drainage connects to the below-ground drainage this should be increased to 450mm to avoid taking the drainage through the foundation.

11.5.1.5.2 Brick or concrete paving

Brick or concrete paving offer an alternative to asphalt. Hand-laying methods suit the restricted headroom, and a light-coloured surface is easy to achieve. Line painting and its maintenance may be reduced by incorporating contrasting paviors.

11.5.1.5.3 Concrete

An *in situ* concrete slab, either groundbearing or suspended, is suitable for slabs where ground conditions are poor or where water levels are high.

In situ concrete also offers good durability and the light-coloured finish is often favoured by clients. *In situ* concrete is also suitable for basements and semi-basements and where high groundwater levels are possible. In these circumstances, issues such as flotation, waterproofing, joints and service penetrations should be carefully considered (Section 11.5.2).

11.5.1.6 Lateral stability

The typical methods of providing lateral stability by the use of cores, shear walls, vertical bracing or moment frames can be adopted in car park structures. The location of the lateral stability systems must consider sightlines, vehicle circulation, natural ventilation and the discontinuities at movement joints.

It is especially important that the vertical elements of the lateral stability system are located appropriately to minimise loads due to high thermal forces, which can be significant if the vertical bracing is located such that it restricts the movement of the structure under thermal actions.

11.5.2 Underground car parks

11.5.2.1 Introduction

The available site area, proximity of adjacent structures, flotation issues, potential flooding due to extreme weather events and methods of construction are the principal design constraints in generating the geometry and structure of underground car parks[74, 76].

11.5.2.2 Categories of underground car park

There are two main categories of underground car park:

Car parks below buildings where the shape of the car park is usually controlled in varying degrees by the shape of the building above; column positions are dictated by integrating the design of the car park with the building above and column sizes are likely to be large. The use of transfer structures, where the geometry of the grid is changed, is usually expensive and not adopted without good reason.

Car parks below open spaces such as roads, squares, sports areas, public parks or similar public access areas. Their geometry is controlled by the method of construction access or by the geometry of circulation. The use of the area above the car park for landscaping, highway or public access infrastructure, will affect the loading, structural form and waterproofing of the car park.

11.5.2.3 Methods of construction

The methods of construction are identical to those for constructing deep basements; i.e., they are critically dependent on soil and groundwater conditions. Where a basement is to be constructed up to the boundary of the site, the space required for the temporary and permanent works may be a prime consideration in the method selected. Sometimes, these methods are directed at avoiding temporary works by using the permanent structure of the finished building to provide temporary support as excavation proceeds. Such methods are generally only used when traditional temporary support methods are precluded by cost or space considerations.

The following methods minimise temporary works and are often used for top-down construction:

- Diaphragm walls
- Contiguous pile walls
- Secant pile walls
- Sheet pile walls

Where space permits and the geotechnical conditions are appropriate, open cut excavations can provide an economical solution.

11.5.2.4 Control of groundwater

As car parks are lighter than other building structures, specific consideration must be given to the method of controlling and managing water ingress and the potential development of uplift pressures. Loads arising from even modest depths of water will give rise to significant effects — particularly in the temporary construction state before full minimum permanent actions are in place — and may give rise to instability if not properly addressed. An uplift pressure of $10kN/m^2$ per metre of unbalanced water depth that can arise above the base level must be considered in both the permanent and temporary load cases. In clay soils, the potential for a link through drainage to remote flood conditions, through fissures to artesian pressures and the possibility of perched water tables must all be considered since they can lead to enhanced uplift pressures.

Using pumped systems or pressure-relief valves to limit uplift pressures often appears to provide initial economy, but will require a constant operational and maintenance commitment throughout the life of the structure to ensure significant uplift pressures cannot be generated. The operational cost, risk and consequences of failure of such systems must be balanced against the initial costs of passive measures through the use of thicker base slabs or tension piles in conjunction with watertight boundaries to overcome potential buoyancy.

Acceptance of the risks of pumping and other alternatives, such as provision of pressure-relief valves to allow controlled flooding of lower levels, and balancing of water levels in extreme circumstances, may be considered acceptable for some car parks but must be agreed in advance with the client and relevant authorities.

Even with watertight boundaries, it is recommended that underground car park levels be equipped with positive drainage systems linked to sumps and pumps to allow surface run-off and wash-down water to be removed. Great care must be taken in detailing service provisions to ensure the watertight perimeter is not compromised by service connections and ducts that could lead to water ingress (Figure 11.14).

During extreme weather events, flooding of underground car parks may occur. The probability of the event, loading regime and potential pressure reversals should be considered in the design of the basement.

It is particularly important to agree the acceptance criteria and required environment for basement and underground car parks with the client, e.g., the level of humidity or condensation that will be acceptable. This agreement is essential before the design of structural and ventilation works takes place and can be an area of significant misunderstanding if issues are not raised and addressed at an early stage.

Figure 11.14: Hydraulic linkage and hydraulic head considerations

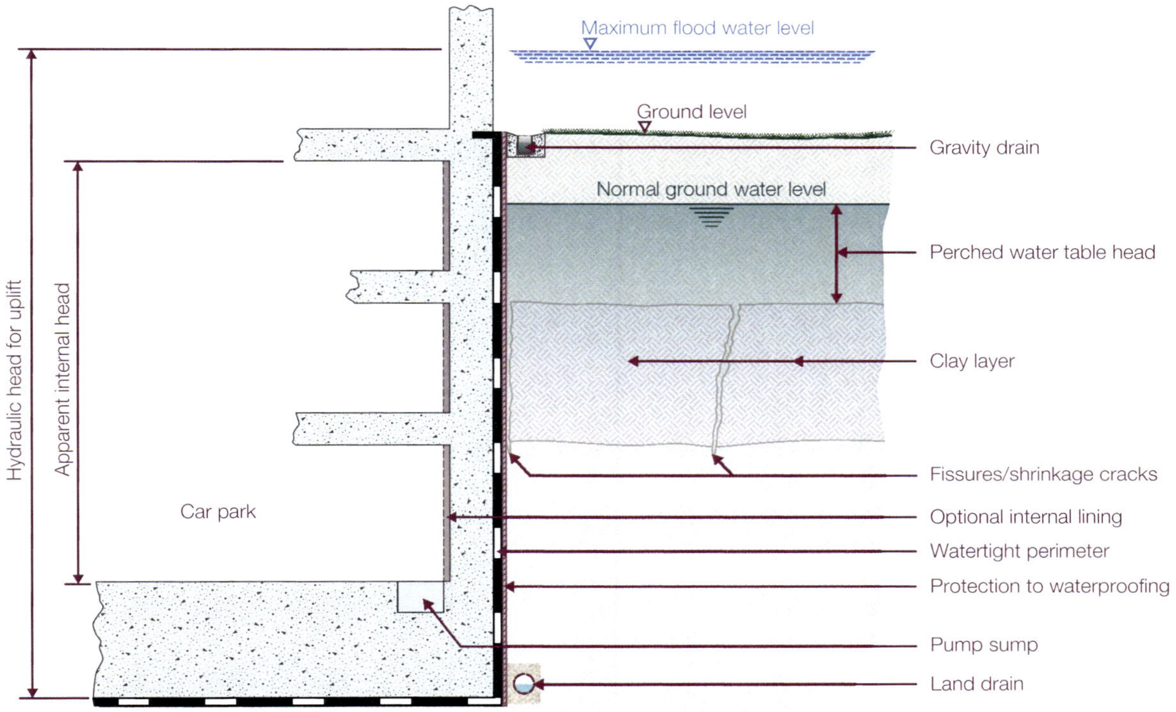

Notes

a. The alignment for approach ramps and flood protection measures may mean flood water levels can exceed adjacent ground levels without flooding the car park entrance ramps.

b. The presence of even small fissures in the clay or less pervious soil will allow the full pressure head to develop, even if flow rates are low.

c. Care is required to ensure service routes do not compromise integrity of perimeter.

11.6 Lean design

11.6.1 Introduction

Car parks are typically functional buildings with unique geometry such as low storey heights, sloping decks, minimal stair cores and a relatively low uniform design load.

The current approach is to design lean for what is needed today, or what we know to be certain in the near future (i.e., where future flexibility is a known, or where a small increase by adding carbon now leads to a huge gain in the future). Also, consideration should be given at the initial design stage to making structures easily adaptable without adding carbon.

Lean design is the practice of utilising fewer resources to make the project more efficient and maximise value. In pure structural terms it is about avoiding overdesign and achieving greater utilisation ratios of the structural elements, including foundations[97].

For car parks that are designed for deconstruction, this is particularly important, and must be considered in order to avoid over-specification of structural materials.

11.6.2 Utilisation of materials

A high degree of material utilisation is fundamental to achieving a lean design. The 2018 MEICON survey[98] identifies the following factors with a negative impact on the final utilisation ratio:

- Ease of construction
- Perceived risk of construction errors
- Lack of penalties for overdesign

Utilisation ratios, including serviceability criteria, should be 90–100% in the final design. If rationalisation of members has taken place, the minimum utilisation ratio should remain above 75%[99].

11.6.3 Substructure

The substructure can account for 20% of the total embodied carbon. The superstructure framing grid should be considered simultaneously with the foundation solution to ensure the final solution is optimised for both structural systems. The superstructure should be light, and where feasible designed to accommodate greater settlements.

Foundations are rarely subject to the full design load until the car park is in use, so the 56-day concrete strength can be utilised rather than the default 28-day strength.

11.7 Deconstruction

11.7.1 Introduction

Circular economy principles require greater resource efficiency which can lead to a reduction of embodied carbon through the reuse and recycling of components and materials from buildings at the end of their useful life. The built environment accounts for approx. 60% of the UK's material consumption and around 50% of all waste arising[100].

Design for Deconstruction (DfD) can realise long-term benefits in the amount of material used, related embodied carbon and reduced waste arising, by ensuring that specified components are fit for purpose and can preferably be reused (rather than recycled) at the end of their life.

11.7.2 General

DfD is an important part of improving levels of resource efficiency and supporting the principles of a circular economy. However, the functional nature of the typical car park structure and its relatively low uniform design load often lead to a lean design approach (Section 11.6) and for car parks that are designed for deconstruction this is particularly important and should be considered to avoid over-specification of structural materials.

Current recycling rates for waste from demolition are very high but the majority of recovered materials are downcycled, resulting in a lower-value product than the original, and lower levels of component reuse. The basis of the design of these buildings and the materials selected will have a significant influence on the rate of reuse.

By carrying out deconstruction, components and materials should be able to be removed for reuse. Demolition on the other hand, often results in materials being destroyed and then recycled. DfD requires two main considerations:

- The nature of materials and components selected
- The means by which the materials and components are connected, and their ability to be deconstructed for reuse

DfD principles need to be considered at the early stages of developing the design philosophy as they will influence both conceptual and detailed design.

Project records are key to realising the benefits of DfD. When considering deconstruction/demolition, the contractor must have sufficient information to identify the components and materials used and their potential for reuse or recycling. Building Information Modelling (BIM) will play a crucial role in delivering DfD, through its ability to provide intelligent data on all the project assets.

11.8 Drainage

11.8.1 Introduction

Investigations have shown that a large proportion of car parks that have deteriorated prior to their first major maintenance schedule have suffered from poor drainage and joint leakage.

If the decks and ramps are not laid to proper falls that encourage water containing de-icing salt to drain away quickly, chloride ions quickly penetrate the surface. Therefore, it is recommended that all decks are provided with provision for surface water drainage to cater for wind-driven rain, water carried in by cars and for periodic washdowns as part of the maintenance regime. If the falls are correct but the gullies designed such that they block easily or are not maintained, a similar outcome is inevitable.

Similarly, joints in the ramps and decks must be properly detailed and maintained, otherwise chloride ion penetration will occur at the slab edges or the cut edges of precast beams. In severe cases, water will run along the underside of the slab, which may have been detailed for a less severe exposure (i.e., it was anticipated that the joint would be maintained and so the beam was not expected to come into contact with de-icing salts) and therefore have less cover to the reinforcement.

Correct specification and maintenance of drainage and jointing systems is therefore extremely important for the long-term durability of the car park.

11.8.2 General

Well designed and properly maintained drainage is essential to ensure that all water deposited on the exposed surfaces is rapidly discharged through an effective drainage system. To achieve this, all drainage outlets should be recessed below the finished top surface of the concrete to ensure that no ponding occurs. Standing water is detrimental to both the structure and the operation of the car park. In the UK, drainage is designed to BS EN 752[101] and BS EN 12056[102]; local standards will apply in other countries. Key issues for consideration, beyond those in the standards, are discussed in the following sections.

11.8.3 Required falls

The specified minimum falls to the finished surface of the roofs and floors are normally based on the:

- Quantity of water likely to fall on the area under consideration
- Texture and accuracy of floor finish
- Sensitivity of structure to deflection and creep
- Anticipated efficiency of any waterproofing

It is recommended that falls in the finished surface should be a minimum of 1:60 at all parking levels — except where particularly smooth, plane surfaces with short drainage lengths can be guaranteed. This minimum requirement applies to all trafficked surfaces, irrespective of whether the area is covered or exposed. The more generous the fall the greater the ability to shed water and to prevent harmful chloride salts saturating the concrete. User comfort also has to be considered and falls greater than 1:20 or sudden changes in fall should generally be avoided. The decks in basement and underground car parks should also have a minimum fall of 1:60 to allow for washing down.

11.8.4 Parking areas

The roof is the most exposed area and the drainage should be designed in accordance with BS EN 12056[103] and the appropriate rainfall intensity. In storm events, some build-up of water is inevitable, and the edges of decks may overflow (despite decks being designed to minimise water discharging over the edge). It is recommended that a roof should be considered above the top parking deck. Intermediate floors are wetted by rain blowing through partly open sides including ramps and by snow and ice melting on parked cars. The quantity of water deposited on intermediate decks from this source is likely to be about 2 litres per parking space per parked vehicle in the UK; this also depends on the average stay, with shopper car parks having short stays and more water, while commuter/business car parks will often have a full day stay per bay.

Proprietary drainage network analysis programs and methods are used to design drainage systems.

Although the direction of fall depends on the geometry of the car park, falls and drainage channels should be designed to suit the type of structure and to take account of any sensitive structural details. Provided the drainage is well designed, it is preferable (but not essential) to lay the fall of the decks outwards towards the exterior of the car park. Perimeter upstands to the edges of decks that are laid to fall outwards are preferred, in order to prevent water and burning fuel runoff in the event of a fire.

Where car wash facilities are to be operated inside car parks, allowance should be made for higher than average discharges of water and salt through drainage channels and associated pipework. If recycled water is to be used, the implications on the durability of the structure need to be carefully considered.

11.8.5 Ramps and circulation areas

Falls towards ramps should be avoided but if the geometry of the car park does not allow this, drainage paths should be intercepted to avoid water discharging onto the ramps, or situated at the bottom to restrict the discharge onto the deck below. Adequate drainage should be provided at the bottom of ramps, particularly at roof and entry levels, to remove water swiftly and prevent ponding and possible freezing.

It is often the case that ramps are finished with an exaggerated tamped finish to increase traction and skid-resistance. This can cause water and melting ice to pond in the small depressions. If a heavily tamped finish is a requirement, it is recommended that chevron or diagonal tamping is used to facilitate the rapid drainage of any water.

11.8.6 Pedestrian areas

Where decks are laid to falls, pedestrian areas should ideally be situated at the higher end of the gradient. Staircases, lifts, etc., should be above general deck level, or discreet drainage should be provided to intercept water flowing towards staircases and lifts.

Drainage channels, slots and gratings, etc., should be dimensioned and detailed to minimise risk to pedestrians tripping and falling and allowing the ingress of chloride-ions into the deck.

Care must be exercised in selecting a drainage channel if the system is cast into a suspended deck as this may provide a path around the drainage channel for chlorides to enter the deck at a point of reduced cover.

11.8.7 Rainwater goods

Rainwater pipes should be arranged to be unobtrusive. Wherever possible, downpipes should be located on the shielding side of the column to avoid traffic impact, and fixed so they do not encroach into the adjacent parking space. Protective hoops or shielding may need to be provided where it is not possible to use the structure to shield the pipes.

It is essential that adequate provision is provided for access to and rodding of the drainage system. Traps for protection against salt, grit, oil and petrol entering the surface water disposal system should be provided in accordance with the requirements of the local authority.

It has been found that plastic rainwater goods can burn through in the event of a fire and allow burning fuel and the fire to spread vertically. Non-combustible materials are preferred.

11.8.8 Petrol interceptors

Interceptors are required to prevent oil and petrol spillages from entering the main surface water drainage. Petrol interceptors should be located externally but if this is not possible, they should be in positions that are easily serviceable without disrupting the operation of the car park. Where deep interceptors are located internally, consideration must be given to access and loading requirements for maintenance vehicles.

11.9 Joints

11.9.1 General

The structural requirements for joints are described in Section 11.3.

In locating and selecting the type of movement joints, account should be taken of the need to maintain drainage falls and to minimise ponding. Ideally, movement joints should run parallel to the drainage falls and preferably be located at the higher end of sloping surfaces. Movement joints should not be situated directly under vehicles because in the event of a fire they can burn through and allow the fire to move vertically and/or allow burning fuel to pass onto the deck below.

11.9.2 Proprietary movement joints

Proprietary movement joints should be suitable for trafficked areas but should not impede traffic flow or impose excessive dynamic loading on the structure. Joints in a waterproofed area should be compatible with the system of waterproofing and should be watertight over their entire length, including the ends.

11.9.3 Sealants

Liquid-applied joint sealants should be compatible with the size of joint, the magnitude of movement under diurnal and seasonal temperature ranges, the dynamic effects of traffic, and any spillages during regular use of the car park. The movement requirements determine the movement accommodation factor (MAF) of the sealant. In general terms, the lowest MAF sealant should be chosen compatible with the predicted movement, as high MAF sealants are soft, do not support edges, are more susceptible to damage (e.g., from stones, etc.) and can stick to car tyres and be pulled out.

For external use (e.g., roofs), where a fully waterproof joint is required, sealants are not recommended without a separate waterstop, and proprietary movement joints are preferred (Section 11.9.2).

Local codes will give suitable joint designs for the range of movement expected; in the UK, design is to BS 6093[88] and sealant should comply with BS 6213[104]. The sealant material should possess the following characteristics:

- The chase for the sealant should be formed with square sides
- The life of the sealant depends on how well it is bonded to the sides of the sealant chase. In practical terms, this means the sealant should bond well to damp concrete, and a suitable primer should be used
- As the joint opens and closes, the sealant must deform in response to the movement without loss of integrity over both the diurnal and seasonal temperature range, but not be too soft and flexible
- A soft backing material (e.g., foam rod) should be used to achieve the correct width:depth ratio (e.g., 1:1) recommended by the sealant manufacturer. Exceeding the recommendations by making the sealant too thin or too thick will reduce the life of the sealant
- The sealant must be durable, as periodic removal and replacement is difficult and expensive, but the life is likely to be less than the design service life of the car park
- Sealants used with waterproof membranes should be designed in conjunction with the waterproofing membrane manufacturer. The contractor laying the membrane should also be responsible for providing any sealants, to avoid contractual difficulties

11.9.4 Construction joints and non-structural cracking

Unless construction joints are adequately reinforced to prevent them opening up due to contraction, it may be necessary to seal the joint, particularly if there is a risk of water ingress and reinforcement corrosion.

Similarly, non-structural cracks can open, particularly in response to creep and temperature changes. Although non-structural cracking is an intrinsic feature of reinforced concrete, it nevertheless needs to be controlled to ensure the design service life of the car park is maintained. The subject of non-structural cracking is complex, and a good summary of the processes is provided by The Concrete Society[81].

The appropriate treatment for the construction joint or crack depends primarily on its width, the range of expected movement in response to temperature and loading, future widening of the crack due to long-term drying shrinkage, and its restraint (e.g., by connection to solid stair or lift shafts). Where there is doubt about the extent of movement, a period of monitoring using crack gauges could be considered. Typical methods for sealing cracks are:

- Where minimal future movement is expected, the crack can be sealed by a rigid structural resin such as an epoxy injection system
- Where slight movement is expected, non-structural foaming resins based on polyurethane resins can be used
- Where large movements are expected, cracks may need to be cut out to form a groove and then filled with sealant (Section 11.9.3)

Using a rigid material (e.g., epoxy resin) in situations where large movements are possible may not be appropriate. It is to be expected that the concrete will crack elsewhere owing to the movement strains, and reinforcement in the slab may redistribute the crack, forming many fine cracks in place of one large one. These can be particularly difficult to seal.

11.10 Durability

11.10.1 General

Car parks are often exposed environments. The design must consider both the structural and durability aspects, considering all relevant parameters. The durability of a structure is often more sensitive to site practice than to the choice of materials used (Chapter 17).

Design of car park structures should consider the following specific exposures, which require more durable grades of concrete than a regular building:

- Effect of prevailing atmospheric exposure conditions, in particular the proximity to coastal areas, where salt deposition onto unprotected flat slabs and other surfaces is a factor
- Use of chloride-based de-icing salt on roads around the car park
- Use of de-icing salt on car park deck surfaces (which is discouraged)

Note, that specific guidelines exist for atmospheric exposure near the coast, as this will be location specific (Section 11.10.2.2).

Recommendations are available on the durability requirements for the three common forms of car park construction:

- Concrete-framed structures and concrete decks and ramps: the durability requirements in Europe for *in situ* and precast concrete are determined from BS EN 206[105], and in the UK BS 8500-1[106], as required by BS EN 1992-1-1[86] (Section 11.10.2)
- Steel-framed structures (including composite steel/concrete decking and steel decks and ramps): the durability requirements are determined from EN ISO 12944-4[107], which gives guidance on the life expectancy of the different paint protection systems (Section 11.10.3)
- Concrete basements and buried structures: determined from BS EN 206 and BS 8500-1 and considering aggressive ground[108] (Section 11.10.4)

In extreme coastal conditions and where de-icing salts are frequently used, additional measures may be needed to enhance the durability of concrete and steel structures. The extent and value of these measures should be considered against the cost and disruption of maintenance and remedial work during the design service life of the structure, as part of whole life asset management (Chapter 18).

11.10.2 Concrete durability

11.10.2.1 Introduction

Away from the foundations, where sulphate risk has to be considered, the main exposure classes for the superstructure are given in BS 8500-1:

- XC: Corrosion induced by carbonation
- XD: Corrosion induced by chlorides other than from seawater
- XS: Corrosion induced by chlorides from seawater
- XF: Freeze-thaw attack

BS EN 206 requires each element or group of elements (columns, parapets, decks, ramps and foundations) to be assigned exposure classes and each given a severity rating (e.g., XC1–XC4), with the higher number indicating greater severity.

The common mechanism for deterioration of concrete above ground is carbonation (XC), leading to corrosion of embedded reinforcement. The open form of most car parks classifies them as XC3/4, and only for dry, enclosed car parks or basement areas would the risk reduce to XC1.

A more severe corrosion mechanism is exposure to chloride salt (Section 11.10.2.2).

Care is required when considering elements exposed to direct wetting. Car park decks and ramps are classified as XF4. Even where protected by a membrane or joint, these can fail, and the consequences should be considered when specifying the concrete.

Less commonly, concrete can be at risk from alkali-silica reaction[109], and the foundations of a car park can be attacked by chemicals in the soils, such as sulfate attack, acidic soils, thaumasite[110], etc. These deterioration mechanisms are explained in specialist guidance[108], and recommendations incorporated into BS 8500-1.

11.10.2.2 Chloride ion attack

This arises from two common sources: road salt brought into the car park decks and ramps on the wheels of vehicles, or spread onto the deck surface by operators in icy weather (XD); and seawater exposure in coastal locations (or where there is groundwater with elevated chloride ion (XS)). Further guidance is given by BRE[111–113].

Structures in coastal locations are known to be at increased risk of sea salt deposition (XS1), which affects the external envelope and roof slab. In Europe, no general guidelines exist for the atmospheric exposure zone distance from the sea, as this is location specific. Salt deposition depends on the strength of prevailing winds and the effects of sheltering by other structures. In salty soils and where permanently immersed in groundwater, the class is XS2, and structures supported above sea level will be XS3.

In winter, salt collects on the road surface outside the car park and the salt is carried onto slabs and ramps in all parts of the car park by the wheels of cars (XD3). Water runoff from cars, due to rain and melting ice and snow, can also carry the salt-contaminated water into all parts of the car park, collecting in areas with poor drainage, and resulting in cyclic wet and dry conditions. Car park operators may use de-icing salts on the slab surfaces to reduce the slipping risk to users, despite advice to the contrary. Prevention of chloride ion penetration into the concrete from these sources must be a primary consideration in car park durability design.

Evaluation of concrete elements in car parks where premature deterioration has occurred, has shown that surfaces or details exposed to de-icing salt need greater protection and/or higher standards than for concrete elsewhere[114].

A primary cause of premature deterioration is water retention on decks, including rough, textured surfaces or in areas of ponding. Unless there is an effective waterproofing membrane on the concrete surface (Section 11.10.6), ponding water will slowly penetrate into the concrete surface. As well as increasing the risk of damage due to freeze-thaw action, the surface water could contain de-icing salts or (in coastal locations) chloride salts from sea spray.

The combination of poor drainage and the lack of beneficial cleaning or rain-wash on the sheltered lower floors, leads to a concentration of chloride salts on the surface as the water evaporates. Cycles of wetting and drying, which are features of car park environments, also accelerate the ingress of chlorides into the concrete surface and into any cracks. If the concrete is detailed to dry rapidly, with adequate falls to shed water from the surface quickly, this will reduce the corrosion risk.

Problems associated with chloride salt attack and corrosion are much less severe where the design and construction incorporates the following:

- Suitable limiting proportions for water, cement type and cement content
- Specified minimum cover to reinforcement
- Thorough compaction of the concrete
- Well-designed and cohesive concrete mix
- Properly designed and sealed movement and construction joints
- Adequate drainage through proper falls
- Good ventilation, allowing slabs and ramps to dry quickly
- Waterproof membrane systems

Water and salts will find accelerated passages through poorly designed and leaking joints and through honeycombed concrete surfaces. High quality concrete construction is therefore the first line of defence against chloride ingress. Water passing through concrete can be highly alkaline and can contain a saturated concentration of calcium hydroxide. When the water evaporates, it leaves behind characteristic white staining, hollow stalactites, and stains on the slab below. Leakage poses a risk to users of the car park and causes damage to paintwork of cars. Good practice is to provide a drip detail to prevent water running back along the soffit from joints in the slab or at the edge parapets. Table 11.2 summarises many of the factors that can create risks of premature deterioration in car parks.

Table 11.2: Risk factors affecting durability

Potential durability problem	Actions to be taken individually or together to minimise problem	Relevant section in this guidance
1. Uncontrolled cracking[a]	• Choice of structural system (e.g., *in situ*, precast, post-tensioned, composite decking)	11.5
	• Design to control crack sizes	11.5
	• Use of synthetic fibres for plastic and drying shrinkage control	11.10.2.4
	• Adopt thorough curing regime	17.2.8
	• Design for movements and volume changes	11.3
	• Select low shrinkage concrete mixes	11.10.2.4
	• Ensure correct finishing practices	11.10.5
	• Correctly detail reinforcement (e.g., staggered laps)	17.2.2
	• Consideration of additional diagonal anti-crack reinforcement at corners of voids and other stress risers	11.5
2. Leaking through slabs	• Detail for crack control	see "Uncontrolled cracking" and 11.10.2
	• Provide good drainage	11.8
	• Properly install and maintain joint systems and liquid-applied sealants	11.9
3. Freeze/thaw damage	• Provide adequate falls and drainage	11.8
	• Specify air entrained concrete	11.10.2.4
	• Check air entrainment level of concrete at discharge	11.10.2.4
	• Use protective coatings and membranes	11.10.6
4. Corrosion of reinforcement	• Assess all relevant exposure conditions	11.10.2.3
	• Utilise concrete mix enhancement	11.10.2.4
	• Adopt corrosion protection measures	11.10.2.5
	• Ensure adequate design of cover to reinforcement	11.10.2.5
	• Use enforcing cover to reinforcement at construction	17.2
	• Provide adequate drainage	11.8
	• Use protective coatings and membranes	11.10.6
	• Ensure only high quality concrete is used	11.10.2
5. Poor concrete mix quality	• Use the correct cement type	11.10.2.4
	• Use a low shrinkage concrete	11.10.2.4
	• Use admixtures to improve placing and minimise water demand	11.10.2.4
	• Use effective curing	17.2.8
	• Use correct finishing practices	11.10.5
	• Enhance the concrete with coatings and membranes	11.10.6

Note:
[a]Cracking can be controlled, but not completely eliminated.

11.10.2.3 Assumptions

Guidance on durability will vary depending on jurisdiction, but the following assumptions are considered implicit in the preparation of a life-care plan:

- De-icing salts are not applied directly to the elements by the car park operator as part of a maintenance regime
- The car park will be well-drained and drainage maintained
- The car park will have good ventilation
- The car park has a design service life of at least 50 years

Any durability assessment also needs to consider that:

- Concrete for exposed top decks and ramps will be subject to rainfall and other forms of wetting and is at risk of freezing while wet, with the risk the membrane may fail at some point
- The same durability requirements should be applied to wearing screeds to precast units, as to structural concrete framing

- Intermediate levels of a car park are often wetted by rainwater; brought in on tyres of vehicles or penetrating through partially-clad sides
- Local conditions will dictate whether higher requirements for freeze-thaw resistance are required on intermediate decks. If frost problems have arisen, they are normally associated with ponding and/or poor-quality concrete. In colder climates, more rigorous measures may be necessary
- Soffits, columns and walls are rarely exposed to splash and spray from de-icing salts, as vehicles move slowly

Extreme exposure situations will increase the risk of premature corrosion of reinforced or prestressed concrete and require special consideration in the following situations:

- Design service life of more than 50 years: carbonation and chloride ingress are time-dependent and extra cover or a higher strength class may be needed
- Structures exposed to tropical or arid climates: higher temperatures will accelerate carbonation and chloride ion penetration
- Underground car parks or foundations exposed to saline groundwater: unless effectively tanked, evaporation of groundwater passing through the walls may concentrate chloride salts at the inner reinforcement layer

11.10.2.4 Specification

The recommendations in Table 11.3 are for car parks in the UK based on BS EN 206 and BS 8500-1 exposure classes, which control among other things the concrete compressive strength, cement type and cover to reinforcement.

The recommendations should provide protection against deterioration, subject to the limitations listed in the assumptions (Section 11.10.2.3). Where the limitations cannot be satisfied, the specification requirements for the concrete will need to go beyond the recommendations in Table 11.3.

Concrete strength and other properties are selected for the worst case combination of exposure classes. For car parks, particular care is needed when considering chloride ion exposure (XD class) and freeze-thaw resistance (XF class). The exposure class will need to be re-assessed if any of the durability assumptions are invalid.

Table 11.3: Recommended exposure classes for use with BS 8500[e]

Element type and location	Recommended exposure class	Recommended exposure class in coastal areas
Top surface of decks and ramps at entry level of car park	XD3 (XC3/4)[a] and XM1/AR4[b]	XD3 (XC3/4)[a], XS1[c] and XM1/AR4[b]
Top surface of decks and ramps exposed to freezing e.g., roof level	XF2 and XD1 (XC3/4)[a] Optional: XM1/AR4[b]	XF2, XS1 (XC3/4)[a] and XD1[d] Optional: XM1/AR4[b]
Top surface of decks and ramps in other locations	XD1 (XC3/4)[a,b]	XS1 (XC3/4)[a] and XD1[b,d]
Soffits of decks and ramps	XC3/4	XS1 (XC3/4)[a]
Vertical elements	XC3/4	XS1 (XC3/4)[a]
Vertical elements exposed to freezing	XC3/4, XF1	XS1 (XC3/4)[a], XF1
Elements protected from rainfall e.g., internal areas such as stair enclosures	XC1	XC1

Notes:

[a]Exposure classes given in brackets denote classes which are less critical and assumed in BS 8500-1 to occur simultaneously with the main exposure class.

[b]While Clause 4.4.1.2 of BS EN1992-1-1 advises that for abrasion class XM1 (moderate) a sacrificial layer of 5mm of concrete may be used, for car parks abrasion class AR4 (light duty industrial and commercial)[115] is recommended as equivalent (Section 11.10.5.2.3 of this guidance).

[c]XD3 condition is more critical.

[d]XS1 condition is more critical, depending on proximity to coast (Section 11.10.2.2).

[e]Alkali-silica reaction, and sulfate or other forms of attack (where present) should be assessed separately to BS 8500-1 requirements.

The exposure classes may be reduced if the surface is to be protected with a waterproofing membrane (Section 11.10.6), but this must be maintained to remain effective; the concrete surface will become vulnerable to chloride ion ingress and freeze-thaw if the membrane fails and is left unrepaired, due to cracks, tears or where the surface has worn away.

By reference to BS 8500-1, either a designed mix can be provided, with defined properties, or a designated mix selected. If air entrainment of the concrete is not used, BS 8500-1 recommends designated concrete where there is a risk of freezing while wet, unless the surface is protected by a liquid-applied waterproof membrane; for example, RC28/35 is suitable for XF1 and RC32/40 for XF2 exposure, otherwise the air-entrained PAV1 mix can be used. With very severe freeze-thaw cycles, the exposure class may need to be increased.

Durability specifications can be enhanced by a combination of measures that either protect the concrete or enhance its penetration resistance to chloride ions and other aggressive agents. Changes in mix composition can significantly improve the durability of the concrete and reduce the risk of premature deterioration[106]. The choice of cement types and aggregate sources should consider the local availability of materials and the ability of the contractor to mix, place, compact and cure the concrete to achieve the intended benefits.

BS EN 206 shows the benefits of using factory pre-blended cements containing various levels of non-Portland materials, including ground granulated blast-furnace slag (GGBS), fly ash (FA) and silica fume (SF). These non-Portland materials can reduce the rate of penetration of chloride ion into concrete for a given strength class, and in the UK, BS 8500-2[13] allows blending of different cements (e.g., CEM I with FA or GGBS) at the mixer in set proportions, and are termed 'combinations'. BS 8500-2 uses a shortened notation to indicate combinations (Table 11.4[106]) (e.g., factory-produced CEM III/B contains CEM I with 66-88% GGBS, and is termed CIIIB in BS 8500-2).

Table 11.4: Common UK cement and combination types

Designation	Notation and composition	Combination
CEM I	Portland cement	CEM I
CEM I-SR	Sulfate-resisting Portland cement	CEM I-SR
CEM II	CEM I + either FA or GGBS or SF in lower dosage (or in combination)	CII
CEM III	Blast-furnace cement: CEM I + GGBS in higher dosage	CIII
CEM IV	Pozzolanic cement: CEM I + FA in higher dosage	CIV
CEM V	Composite cement: CEM I + combination of FA, GGBS, SF	–

Other options for improving the durability of reinforced concrete are summarised as follows[116]:

- Plasticising or superplasticising admixtures: allow water/cement (w/c) ratio to be reduced, improving durability of concrete and resistance to chloride ingress
- Low water demand mixes (based on minimum voidage), tightly packed aggregate of good grading and shape: helps lower w/c ratio and reduce shrinkage of concrete
- Water-resisting admixtures: reduce the absorption of water and water-borne salts into concrete and have been used successfully for waterproof basement construction
- Class I Micro (monofilament) polymer fibres[117]: reduce the risk of plastic shrinkage cracking and improve mix cohesiveness, particularly for flat slab construction exposed to sun and wind
- Steel fibres[118] and Class II Macro synthetic polymer fibres[117]: have the potential to improve the post-cracking properties of hardened concrete and can be used as a corrosion-free alternative to nominal bar or fabric reinforcement[119]

11.10.2.5 Corrosion prevention

Corrosion of steel reinforcement or other embedded metal is a serious problem that may weaken the car park before it has reached its design service life. While modifying or protecting the concrete will achieve some improvement in durability (Section 11.10.2.4), other designs can also be considered for the reinforcement. Alternatively, non-ferrous reinforcement can be used (Section 11.10.2.6)

11.10.2.5.1 Cover to reinforcement

When specifying concrete cover, several aspects must be considered:

- Maintaining the correct cover to the top reinforcement in slabs and ramps, using rigid chairs etc., is an essential step as this is the most critical zone for chloride ion penetration
- Care is required in detailing, to ensure tolerances for placing reinforcement and formwork construction. Concrete thickness and finish are sometimes not compatible with the cover tolerance
- The mass of wet concrete will cause deflections in conventional formwork, which can reduce concrete slab levels and affect cover depths; this can also be a problem with wearing screeds applied over prestressed precast concrete planks or where steel decking is used (Sections 11.10.3.2 and 11.5.1.2.2)
- Cover over column and wall reinforcement should be similar to top cover in adjacent slabs, particularly if these are at or near gutter lines or in areas exposed to salt-laden slush and splash
- Precast members exposed to salts should either include cover requirements at their ends or have equivalent protection (e.g., an impervious coating system), particularly as joints formed above these ends may leak chloride-laden water
- Anchorages and vent tubes for grouted post-tensioned elements are potential locations for ingress of water containing de-icing salt, which could lead to corrosion of the tendons. These vulnerable details must be sealed to protect them effectively and must be regularly inspected
- Cutting or forming of chases or holes in structural members must not be undertaken without due consideration of the structural and durability implications

11.10.2.5.2 Corrosion-resisting steel reinforcement

In addition to high-quality concrete cover, additional protection of the reinforcement can be provided by the choice of steel or by using coatings.

Austenitic 316 stainless steel minimises the risk of corrosion in concrete[120]. Due to its cost, it is usually only appropriate in local areas where conditions are particularly severe, e.g., barrier fixings and bearing shelves. It should be detailed to ensure that it is electrically isolated from other steel reinforcement; electrical contact between stainless and conventional reinforcing steel will induce galvanic corrosion.

Galvanised reinforcing steel can extend the time to corrosion of reinforcement in carbonated concrete, but is not recommended for exposure to heavy chloride-contamination found in car parks[121].

Epoxy-coated reinforcement to BS ISO 14654[122] offers improved protection against reinforcement corrosion due to carbonation and chloride ingress. Particular care is needed during delivery, bending, cutting and site placement, to repair any exposed areas of steel which would otherwise lead to localised corrosion.

11.10.2.5.3 Protecting prestressing strand

The higher quality normally used in precast prestressed concrete construction, including higher strength, less permeable concrete and higher precision on cover, has resulted in fewer early problems with prestressed units compared with *in situ* concrete.

Prestressed precast planks are usually finished with a high strength wearing screed, to BS 8204-2[115]. A range of problems associated with cracking and movement of the screed above the prestressed units can result in leakage. Problems can also occur at the ends of the units, particularly at bearing shelves, where leakage through joints can penetrate into the prestressing curtailment area or cling to the soffits, leading to corrosion of the strand.

The highly stressed strand is more sensitive to loss of section than conventional reinforcement, which can lead to sudden, undetected failure of strand wires. It is particularly important to detail and protect these units, especially at the cut end faces, and inspect the vulnerable units regularly. The potential problems from the penetration of chloride-bearing water into hollow precast units also need to be considered when detailing drainage.

Post-tensioned tendons can be provided in various shapes and forms, and comprise two main types:

- Fully-bonded (grouted) tendons in ducts or in pre-formed holes (ductless)
- Unbonded tendons, normally in polypropylene sleeves encased in corrosion-inhibiting grease

Compared with pre-tensioned tendons, post-tensioned design allows the tendon depth to be varied along the element. Problems have been found with corrosion of grouted post-tensioned tendons inside ducts, particularly in bridges[123]. Research findings highlight the importance of using pressure-tested ducts, appropriate grout, a high standard of grouting practice and checks on the quality of the final grout and effective protection of anchorages by waterproof mortar and/or protective coatings.

11.10.2.5.4 Other embedded metals

Electrical contact between dissimilar metals, particularly between uncoated reinforcement and stainless steel, brass, bronze, copper (also used as flashing material) and lead (sometimes used around drains) should be avoided. These metals can promote galvanic corrosion of reinforcing steel if they are in electrical contact with it, and the surrounding concrete becomes contaminated with chloride ions. Particular care is needed with the detailing of the many fixings into the concrete that are often required in car parks. It is also important to isolate galvanised or aluminium elements from non-galvanised reinforcement, as the galvanised coating or aluminium element will corrode sacrificially.

Embedded metal electrical conduit can adversely affect the structural performance of concrete, as ducts may become a route for chloride ion ingress. Steel conduit with insufficient cover can rust, and unprotected aluminium conduit can be susceptible to severe corrosion in moist concrete. It is recommended to fix metal conduit to the concrete surface, rather than embed it. Embedded plastic conduit should be used with caution as it can form a path for water and chloride ingress, and can lead to cracking in areas where the concrete cover has been reduced by the conduit.

11.10.2.5.5 Controlling electrochemical corrosion

Cathodic protection (CP) of reinforced concrete is being used increasingly for new-build structures in extreme exposure conditions to prevent corrosion; including bridges, tunnels and other structures with long service lives exposed to seawater conditions, and for car park structures in saline ground. It is also used frequently for the repair of existing structures, including car parks.

The performance requirements of CP systems are set out in BS EN 12696[124], with further guidance and interpretation given in TR73[120].

11.10.2.6 Non-ferrous reinforcement

Unlike steel reinforcement, fibre polymer composite (FPC) bars do not require a cement-rich, highly alkaline environment for their long-term corrosion protection. Quality-controlled FPC bars are now available as a non-ferrous alternative to steel bars. Early experience indicates that these materials offer high durability in aggressive chemical environments, including those experiencing salt and fuel contamination[125].

11.10.3 Structural steel durability

Steel may be used as an alternative to reinforced concrete for framing in composite design or for concrete-filled steel columns. All-steel framing and decks are also used, particularly for demountable car park construction.

11.10.3.1 Corrosion protection

Guidance on corrosion protection for structural steel above ground can be found in BS EN ISO 12944-4[107]. Proximity to the coast is much more significant for steel and galvanised structures, and depending on the prevailing conditions, additional protection may be required. The structures are classified by six atmospheric corrosivity categories, assessed either by standard exposure specimens, or by considering yearly time of wetness, yearly mean concentration of sulphur dioxide and yearly mean deposition of chloride. No specific guidance is given for protection of structural steel for use in car parks subject to local splash and spray from cars and broadcast de-icing salt.

Where exposure to chloride ions is expected, either due to run-off or direct exposure from sea spray, the car park durability design needs to address this enhanced corrosion risk. In particular, consideration of the selection of steel components and detailing is required, to provide appropriate corrosion protection and to avoid mixing different types of steel or metals leading to bi-metallic corrosion.

Primary requirements of the design are to prevent exposure of the steel frame and steel fixings to direct/indirect salt exposure and impact damage and to meet fire requirements. Where these have been achieved, regular steelwork painting systems for outdoor exposure can be used, subject to consideration of fire protection.

Where there is a risk of direct salt exposure e.g., at the base of steel columns adjacent to driveway areas, the higher standard of paint specification used for bridge steelwork may be more appropriate, as set out in BS EN ISO 12944-4.

In extreme exposure conditions, the life expectancy of the paint protection system may be shorter than the design service life of the car park, and programmes of planned maintenance painting may be required at deteriorating areas identified from inspections (Section 18.3).

Particular problems can arise where steel elements enter concrete, creating an interface that cannot be inspected easily, if at all. Steel corrosion can be particularly severe where cracking develops in the concrete at the junction, creating a downward crevice into which organic matter and/or salts can concentrate. This maintains the damp conditions that accelerate corrosion of bare steel. Corrosion can also arise with fixings into holes with cavities or with poor-quality mortar infill that rapidly carbonates. Also, as the surface layers of concrete become carbonated over time, they lose the alkalinity that protects the steel from corrosion.

Corrosion protection on embedded steel must be continued into concrete for at least the specified depth of cover for reinforcement. Slots (typically 10mm wide) should also be provided in the concrete at the junction with the steelwork to receive mastic sealant to BS 6213[104], which should be regularly inspected and maintained.

Bolts into the slab or columns for barriers and cladding fixings, etc. are similarly at risk. The choice of details for barrier and cladding fixings should take into account the ease with which such elements can be removed for inspection and replacement if there is a risk of deterioration.

11.10.3.2 Composite concrete and steel decking

Profiled galvanised steel sheeting is often used as permanent soffit formwork for reinforced concrete slabs which act compositely to provide external tensile reinforcement for the slab. Design of this form of construction is comprehensively covered in TR75[126]. It is also used for car park decks and ramps, but has to be used with care. The concrete used with steel decking is relatively thin and can be expected to crack. Where this occurs, water and salt will penetrate through the surface and down to the galvanised steel sheeting, incurring a risk of corrosion (Section 11.10.3.2.1).

Experience has shown that these systems can have a reduced service life, generally estimated to be around 30 years. Therefore, it is unlikely a 50-year life will be achieved unless an effective waterproofing system is used and is regularly inspected and maintained. Particular care is required in providing good falls for drainage, and sealing all cracks before applying a reinforced waterproofing system to the concrete (Section 11.10.6).

11.10.3.2.1 Corrosion risk

Metal decking should be used as permanent formwork only, i.e., not as external tensile reinforcement as it is with office floor design to TR75. Even if it is used non-structurally, corrosion of the steel decking may still be unacceptable as a service criterion, and therefore it should only be used if it has been appropriately protected and will be properly maintained.

The variable actions (i.e., traffic) and low concrete thickness over ribs both give rise to a greater risk of cracking in the lightweight construction. In conventional applications of composite decking in the dry environments of offices and other building applications, cracking would be of little consequence beneath floor finishes.

With car parks built in areas where de-icing salts are used, or where there are other chloride sources, the water run-off can be carried through cracks in the concrete slab, and become trapped between the concrete and the profiled steel sheet, leading to corrosion. Corrosion damage on the upper, hidden surface of the steel is very difficult to detect without physical examination, and unexpected corrosion may reduce the life expectancy of the slab before major maintenance is required.

For these reasons, additional attention to detail is needed when metal decking is used for car parks, and good drainage measures must be used to shed water from the surface. Any cracking should be sealed, and an elastomeric waterproof membrane applied, reinforced over cracks to prevent water penetration (Section 11.10.6). For continued satisfactory performance, the membrane will need to be inspected and repaired or replaced promptly if defects form.

Some permanent metal formwork systems have a special coating on the inside to enhance their corrosion resistance. There is no data currently available demonstrating how long the coating prolongs the service life of the car park, and so these systems should not be used without an effective waterproof membrane on the concrete.

Another corrosion issue is at the metal-to-metal joint between the galvanised decking and the top flange of the structural framing. Commonly, the steelwork is only provided with a shop primer and has less resistance to corrosion than either a full paint system, or a galvanised surface or concrete-to-steel contact where the concrete offers some passivity. The structural metal sheets are usually corrugated, drawing rainwater and condensation down to the top flange. In coastal areas, wind-driven salt and spray will also get into the crevice between the top flange and the decking.

11.10.3.3 Fire protection
Extra considerations are required for steel elements that are likely to become wet in service from rainfall or splash, which will compromise certain fire protection systems and/or lead to a build-up of moisture and chloride salts in the fire protection, presenting a possible corrosion risk to the steel (Section 11.10.6.4.3).

11.10.4 Basements and buried structures
The deterioration processes for car park ramps and decks built below ground are similar to those for above-ground structures described in Section 11.10.1, subject to the following specific considerations:

- Soil and groundwater conditions may pose a risk of sulfate attack on the concrete
- Brackish or saline groundwater can be concentrated at the inside face of reinforced concrete walls and base slabs, leading to chloride ion attack of the reinforcement
- Bacterially active clay soils can form sulfate and acid by bacterial interaction with harmless sulfides, and enhanced protection may be required

Reference to standard durability provisions in BS 8500-1 provides the necessary guidance on how to design for these specific forms of exposure.

The Concrete Centre provides comprehensive guidance on various aspects of below-ground concrete construction[127].

11.10.5 Concrete finishes
11.10.5.1 General
The floor of a car park should be serviceable but not noticeable, and resistant to wear from wheeled traffic. Puddles, crude irregularities, cracks and stains will attract the attention of the driver and passengers. Puddles of water that contain chloride ions from de-icing salts will also pose a threat to the durability of the structure unless membranes are used (Section 11.10.6). The appearance of the car park deck and general decor and lighting can have an important influence on the customer's reaction to a car park, compared with an industrial hardstanding or road surface[119].

Uneven, heavily tamped surface finishes will be uncomfortable to walk on in soft-soled shoes and make manoeuvring trolleys more difficult, yet a smooth floated surface has less skid-resistance than a textured finish. Power-trowelled surfaces are exceptionally smooth and of low skid-resistance, and also increase the levels of tyre noise in turning areas. The surface should also be suitable for the application of traffic direction and bay markings. Vehicle speeds in car parks should be low and, even in the wet, skid-resistance may not be as critical as for standard highway design except for exposed surfaces subject to freezing.

Surfaces of consistently good appearance can only be achieved with consistent materials, timing and surface finishing processes. Even minor changes in day-to-day methods cause irregular textures and patterning that detract from the appearance of a floor.

A concrete deck not intended to be waterproofed with a membrane should be designed and detailed, specified and constructed to drain effectively, thereby preventing ponding of water on the surface. The positions of construction joints need to be well planned, as they are the points of weakness where leaks may occur. It is prudent to locate vulnerable details such as construction joints away from areas likely to be subject to water collection and run-off. This approach assumes that decks between planned joints are uncracked and that any construction joints that open up because of shrinkage or temperature change will be sealed at a later time, as part of routine maintenance.

Unplanned non-structural cracking of car park decks is a common problem where:

- Plastic cracks form in the first few hours after casting. These can be prevented by good site practice and use of Class I microfibres (Section 11.10.2.4)

- Thermal cracking is caused by restraint; either external (e.g., from an adjacent concrete pour or a base) or internal (where a temperature differential of ≥20°C develops between the cool surface temperature and the centre of large concrete pours)
- Restraint to drying shrinkage causes cracking during the first two years of service

Careful, conservative design is needed to allow for drying shrinkage, creep and thermal movements and to prevent unplanned restraint. A common problem is caused by fixity of the car park deck to rigid stair cores at either end, which restrain movement and cause unplanned cracking. Proper design for movement is essential[83].

Where significant cracks (greater than 0.3mm) occur in the concrete deck or ramp, these should be sealed to prevent water penetration. If the crack is measured and shown to be passive, such that no future movement is expected, repair can be made using a structural resin certified to BS EN 1504-5[128] that will transmit force (Category F). However, where movement is reversible or seasonal, care must be taken to ensure the crack will not simply re-form alongside the re-bonded crack. A ductile resin (Category D) or swelling and foaming resin (Category S) should be used. Alternatively, where the crack is relatively linear, a mastic-sealed formed joint may be cut into the slab, by saw-cutting a parallel-sided chase over the crack.

BS EN 1504[129] is a performance standard for products and systems used in construction, including new-build as well as repair works. Products and systems complying with the standard receive a CE/UKCA mark certifying among other things that; it meets the performance requirements, it has an approved quality assurance scheme in place and it meets regulations for health and safety and the environment. Note, that from 1 January 2025 CE marking will be replaced with UK conformity assessed (UKCA) marking in England, Scotland and Wales. Workmanship on site for the preparation and use of injection resins is set out in BS EN 1504-10[130].

11.10.5.2 Floor finishes

11.10.5.2.1 General

Even when water only occasionally ponds on intermediate floors, blotchy, dusty patches can result after drying out. In the direction of fall, a standard of regularity at least equal to that of concrete floor screeds to SR2 given in BS 8204-2[115] is recommended: similar regularity tolerances are given in BS EN 13670[131] and the NSCS[132].
BS 8204 requires a maximum permissible departure from a 2m straight-edge resting in contact with the floor to be less than 5mm. NSCS indicates a global deviation of 9mm from datum.

Attaining an acceptably smooth surface is not a separate stage in the floor casting process. Success comes from carefully setting out the intended finished levels, placing concrete evenly, compacting it uniformly and controlling the amount of surcharge ahead of the straight-edge when striking-off. Regardless of the texture specified, finishing should be a separate process that follows striking-off.

Abrupt irregularities and depressions of any depth over 3mm should be avoided, particularly where a waterproof membrane is to be applied (Section 11.10.6). Regularity is largely determined at the striking-off stage, since the finishing techniques merely change the character of the upper surface and remove minor imperfections in the surface, and it is too stiff at this stage for changes to be made.

Special care is needed with the levelling and finishing of certain types of construction if ponding is to be avoided, and mid-span propping may be necessary:

- Composite concrete and steel decking can be prone to mid-point defection as concrete is added[126]
- Prestressed precast concrete planks can also deflect as the wearing screed is added and the pre-camber at mid-span reduces
- Longer-term creep and shrinkage can increase mid-span deflections over time

Table 11.5 (based on BS EN 13670 and NSCS guidance) provides advice on executing concrete works, and discusses types of finish. No specific guidance is given for the unformed surface finish of car park decks and ramps, and these need to be considered carefully by the designer, with trial areas specified as required to ensure uniformity. The traditional standard for car park surfaces that do not require membranes or other finishes is recommended as an 'Ordinary' finish or a 'Special (Textured)' finish.

11.10.5.2.2 Finishes for car park deck and ramp surfaces

The following recommendations are for unformed (UF) surfaces using the Type definitions in Table 11.5.

Table 11.5: Concrete finishes and common applications

Type	Application	Examples
Formed surfaces		
Basic finish	Where no particular requirement is needed.	Foundations.
Ordinary finish	Where not of visual importance or to receive applied finishes.	Areas with applied render finish or unseen surfaces such as inside ducts or lift shafts.
Plain finish	Where visual effect is of some importance.	Areas seen occasionally and areas with prepared, direct painted zones where there are some particular requirements.
Special finish	Where special requirements are appropriate.	Areas where surface regularity and/or colour are important.
Unformed surfaces		
Basic finish	Surface levelled by skip float or similar process to produce a closed uniform surface. No further work is required.	Areas to receive a screed finish or other applied finish. Float marks and ridges will occur.
Ordinary finish	Surface finished by floating or panning or similar process to produce a level, uniform surface. No further work is required.	The finish produced has a fine texture and is suitable for trafficking, although the surface will wear. Also for areas receiving a membrane.
Plain finish	Surface finished by trowelling or similar process to produce a dense, smooth, level, uniform surface. No further work is required.	This finish is hard, dense and smooth but provides minimal slip resistance when wet.
Special finish	A surface where special requirements are appropriate for further working of another finish.	This finish should be specified where a defined texture is required. Examples are brush, turf drag, hessian, tine or tamped[119] (also refer to Section 11.10.5.2.2 of this guidance).

Notes:
Derived/adapted from the NSCS and Table F4 of BS EN 13670.
A 'Special' finish does not necessarily mean 'high quality'. It is a method for specifying and agreeing aspects of a surface finish which are outside the descriptions given for 'Basic', 'Ordinary' and 'Plain' finishes.

Special (Plain finish, unpolished)
This provides a smooth but unpolished surface, sometimes called a 'matt trowelled' finish. Usually required only in areas where waterproofing is to be applied or for forming water-collecting channels. The floor smoothness should be appropriate to the proposed type of waterproof membrane. For a uniformly smooth surface, a power trowel technique is often used but with minimal passes so that polishing does not occur. The exact requirement should be checked with the waterproof membrane manufacturer. Because this surface is usually prepared by vacuum shot blasting before applying the membrane, achieving perfection in the finished surface is not as important as for a 'Special (Plain finish, polished)' surface.

Special (Plain finish, polished)
Power trowelling after floating quickly produces a dense, smooth and hardwearing surface with negligible 'ripple' or 'chatter' marks from the trowel. The surface not only has advantages in terms of ease of drainage and reduced water retention, but also improves the wear and durability of the surface, minimising the risk of dusting. The surface has a much lower skid-resistance than a textured finish, especially when the surface is wet. The surface also causes more squeals from rubber tyres than other finishes.

Special (Textured finish)
This provides a fine texture onto a smoothed concrete surface that has been floated to the correct regularity and falls. Texturing improves the appearance and enhances skid-resistance. Texturing can be applied using a variety

of methods once the concrete has begun to stiffen. Common examples include brushed finish or tine finish. Other examples are given in TR66[119]. These finishes can suffer from variations in uniformity from pour to pour and unless specified carefully can adversely affect free drainage.

Special (Grinding)
Grinding can be used as an alternative to wet trowelling or to rectify an unsuitable surface. The machine skims the concrete to expose a sound, dust-free and hard surface. Grinding cannot easily correct bad surface regularity, which may require a concrete plane, but can improve dusting from a surface that was poorly cured. Before grinding or planing the surface, the effect on the cover to the reinforcement should be checked and if necessary, alternative measures considered; e.g., applying a waterproof membrane (Section 11.10.6). Unless combined with vacuum collection, grinding is a dusty process. The cost is reduced when grinding is carried out 24–48 hours after casting, when the concrete is still relatively weak. The smooth finish that results will have similar disadvantages to the 'Special (Plain finish, polished)' finish in terms of skid-resistance and tyre squeal.

Special (Tamped surfaces)
Although previously popular, the use of a tamped finish is now less common for parking areas used by pedestrians. The use of large area pour methods for slabs also makes it difficult to provide a uniform tamped finish, as there is no guide rail against which the concrete can be worked. These finishes are normally used for vehicle ramps and circulation areas.

A 'Special (Tamped)' finish is produced by raising and lowering a compacting beam in its final pass to produce a surface with ridges at a fairly regular spacing (20–30mm) with a height of up to 5mm. Heavier tamping may produce deeper ridge depths, but is not commonly used for car parks. It is difficult to maintain an even distribution of ridges and uniformity from pour to pour, due to slight changes in the concrete mix and/or personnel. Also the terms 'light', 'medium' or 'heavy' tamp are hard to quantify in a specification. A trial panel is recommended as a control for the actual works.

11.10.5.2.3 Abrasion resistance
Resistance of the floor to wear and abrasion will affect the appearance and functionality of the ramps and circulation areas. Unless abrasion of the surface is considered, uneven wear will occur, with weak, laitance-rich areas becoming quickly eroded, leading to depressions that will themselves lead to other problems including ponding of water on the surface.

Little specific guidance exists on how to deal with surface abrasion and wear in car parks. General guidance on abrasion and wear is given in BS 8204-2[115] for direct finished (DF) concrete and concrete wearing screeds. The standard refers to BS EN 13892-4[133] which describes a test to measure the abrasion resistance of a floor surface based on wear caused by steel wheels. It identifies the benefits that can be achieved from upgrading the surface by adding a dry shake/sprinkle finish into the setting concrete and then trowelling this into the surface.

Direct finished surfaces on decks and ramps will receive moderate abrasion from rubber-tyred traffic. According to BS EN 13892-4, the approprite surface class in Tables 3 and 4 is AR4/DF (light duty industrial and commercial). When tested immediately after curing, the surface should have a maximum wear depth of 0.4mm using the standard test method.

BS 8204-2 also covers the abrasion resistance of wearing screeds applied, for example, over precast concrete planks, which commonly also contribute to the structural capacity; BS EN 1992-1-1[86] refers to this as a 'structural topping' rather than a wearing screed, although the workmanship recommendations of BS 8204-2 can also be applied. The same abrasion resistance can be used as that for DF concrete (AR4/WS).

These recommendations assume the concrete surface is a smooth trowelled finish, and so the assessment method is not appropriate for ramps or slabs of substantial slope, and is usually unsuitable for the assessment of Special finishes such as 'Tamped' or 'Textured'.

After hardening, floor hardeners can be applied to reduce surface dusting and improve abrasion resistance. A suitably constructed and well cured car park slab or ramp should not require a floor-hardening treatment. The occurrence of dusting depends on the effectiveness of the initial curing of the slab and the severity of the abrasion. In general, more dusting can be tolerated in a multi-storey car park than in most factory buildings, although this will depend on the maintenance programme as well as the standard for dust prevention set by the car park owner.

As an alternative to higher strength concrete or floor hardeners, BS EN 1992-1-1 advises that for abrasion class XM1/AR4 (moderate) a sacrificial layer of 5mm of concrete may be used. This approach could lead to problems of ponding on surfaces with only gentle falls as mentioned previously, and is not commonly used for car park structures.

11.10.5.3 Parking and pedestrian areas

With the exception of vehicular ramps, all parts of the car park should be suitable for both vehicles and pedestrian use. All designated pedestrian areas should be clearly defined and any finishes should minimise the risk of slipping; for example, some waterproof membranes and surface coatings can be slippery when wet. The most common specification for these areas is Special (Textured finish).

11.10.5.4 Vehicle ramps and circulation areas

Where vehicle ramps are steeper than 1:10, a Special (Tamped) surface is recommended, with the grooves in a chevron pattern to facilitate drainage. The Special (Textured), Special (Plain, unpolished), and Special (Plain, polished) finishes may not have sufficient traction in wet or icy conditions.

If the tamped finish is too heavy, it will impede drainage and lead to contaminants being trapped in the bottom of the grooves. Because of the lack of compaction in the ridges and the tendency for some residual bleed, this finish can be dusty and will wear unevenly.

Where vehicular ramp and deck slopes are less than 1:10, a Special (Textured) finish may be used.

A trial panel is recommended as a control for the works.

11.10.5.5 Walls, columns and soffits

Smooth, high-quality plain-finished concrete is most suitable for walls, columns and soffits in visible areas and can be specified using Table 11.5. In modern car parks, the surface is often given a decorative or anti-graffiti treatment to improve appearance (Section 11.10.6.4.1).

Exposed edges of concrete sections should be chamfered to enhance their appearance and to improve safety. Although cast-in galvanized or steel corner guards can do much to protect concrete columns in vulnerable locations, careful detailing is required — and unless radiused angles are used, the steel corners can be sharp and pose a risk to persons falling. For this reason, their use should be limited.

Special 'feature finishes' can be used to good effect and can be varied throughout the car park. They should not be abrasive or endanger users. Where deep profiles are used, the design and installation must be carefully checked to ensure the minimum cover to the reinforcement is maintained.

11.10.5.6 Basements and buried structures

In general, the finish is similar to that described for walls, columns and soffits (Section 11.10.5.5) but a special feature finish is often used for effect or for directional assistance, e.g., chevrons cast into the surface.

Special care is required when selecting paint systems for external walls, owing to the risk of higher moisture levels in concrete earth-retaining walls (Section 11.10.6.4).

11.10.6 Membranes and coatings for concrete

11.10.6.1 General

Paints and coatings for buildings and car parks are either decorative or protective. Requirements for application of decorative coatings for steel, concrete and other materials are given in BS 6150[134]. Protective coatings for concrete that are required to resist carbonation or chloride penetration, are termed surface protection systems (SPSs), and are specified through BS EN 1504-2[135].

In car parks, SPSs are beneficial when applied to the bottom 1m of walls and columns to prevent penetration of salty water that may be splashed up by passing vehicles. SPSs are classified into three generic types: 'hydrophobic impregnation'; 'impregnation'; and a 'film-forming coating' that sits on the surface.

A properly applied and maintained membrane for decks and ramps can be a highly effective means of reducing the risk of chloride-induced reinforcement corrosion providing it is maintained over the life of the car park. For this reason, membranes are being used increasingly for entrance ramps and slabs, for roof levels and also for intermediate levels.

Membranes fall into two main types: liquid-applied polymer systems complying with BS EN 1504-2 that are applied by spray or roller and bond onto the concrete surface; and conventional mastic asphalt[136] that has a separation layer between itself and the concrete.

Traditionally, mastic asphalt has only been used on the roof deck and ramps, where full waterproofing characteristics are required. Mastic asphalt is particularly suitable for external application, and will tolerate damp concrete, high humidity and offer early rain damage resistance during application, as well as having in-service performance that tolerates freezing conditions in winter and high solar gain in summer. Many liquid-applied membranes are now available that also meet these performance requirements.

Increasingly, liquid-applied polymer systems are being used on lower levels, particularly entrance areas where salts are brought in from the highway. Intermediate-level membranes have a lower performance requirement and are often thinner, less flexible and more vulnerable to traffic wear and damage than those used on the exposed roof decks. If intermediate-level membranes are used, it is prudent not to reduce the concrete durability exposure class significantly, if at all, in the event the membrane fails.

Decisions on the use of membranes and coatings will usually be based on a cost/benefit comparison, but less obvious factors include:

- When a bonded membrane of coating is required, the type of curing agent to be used should be carefully considered for compatibility with the chosen systems, to ensure strong bond is achieved
- At application, the concrete should not have excess moisture in the surface, as this can cause membranes and coatings to blister. Providing rain protection and allowing adequate drying times can affect the construction programme
- Spray-on membranes require more preparation, a drier concrete surface and more coats than a mastic asphalt, but are likely to be more attractive and can last longer
- Solar radiation, de-icing salts, fuel and oils, shrinkage hardening, and embrittlement all reduce the life of membranes and coatings
- Local defects or cracks in membranes and coatings can allow unexpected chloride ion ingress that, over time, can cause premature corrosion of the reinforcement
- It is essential to maintain membrane integrity at all times, through proper maintenance if their full value is to be realised, and not to downgrade the exposure class[106] in case of such failure
- Re-application, in whole or in part, may be required every 10–20 years, leading to a potential loss of revenue due to the works and closure of sections of the car park
- Thinner 'intermediate deck' membrane systems have a shorter life than thicker roof membranes, and have been found to fail within five years of installation
- The relatively short life and high maintenance cost of top deck waterproofing systems should be compared with installing a traditional roof structure to achieve effective drainage[29]

These points should be discussed with the client and a strategy agreed, as they are long-term considerations.

11.10.6.2 Deck waterproofing (general)

Water leaking through cracks and failed joints can result in damage to car paintwork, particularly when aggravated by local ponding above. Water passing through cracks in the structure or around features such as holding-down bolts for vehicular restraint barriers, drainage outlets, etc., will lead to rapid deterioration of the structure. Ponding of water may also result in a health and safety issue if it freezes, causing cars to skid or pedestrians to slip, leading operators to add their own de-icing salt to the decks and ramps.

Special consideration must be given to waterproofing a car park deck that forms the roof to shops or commercial premises. The cost of eradicating water leaks when the car park is in use usually greatly exceeds the preventive expenditure required at construction stage. Preferably, the membrane should be installed towards the end of construction, protected and certified as compliant with the warranty.

Where car wash facilities are to be operated inside car parks, decks and joints should be waterproofed and extra provision made for drainage. For any waterproof membrane to be fully effective, a carefully developed and effective drainage scheme, particularly for exposed decks, is required (Section 6.3.3).

Prior to the application of a membrane, the concrete surface must be checked and approved as suitable to receive the system. Correct falls must be provided and areas of potential ponding rectified by compatible surfacing treatments or grinding down of high points. Car park deck membranes are not usually suitable for permanent immersion conditions, and long exposure to immersion can result in premature failure.

Good preparation of the concrete is an essential prerequisite to providing lasting membrane protection. At non-structural cracks or at construction joints, substantial movement can occur, as the car park responds to temperature changes, typically moving by 0.5–1mm. Long-term shrinkage of the concrete can also cause cracks to open further, and structural flexure of the slab may cause rapid opening and closing.

With liquid polymer systems, it is unlikely a bonded waterproof membrane will be able to bridge over or withstand substantial movement of non-structural cracks unless it is specially reinforced at these locations. Unbonded mastic asphalt construction may also fail over wide cracks, particularly if there is reversible movement.

Car park deck membranes can usually accommodate fine cracks (<0.3mm) without the crack needing to be sealed. Wider cracks may need to be sealed with a resin system complying with BS EN 1504-5[128] before the membrane is applied, depending on the system used and whether it is reinforced with fabric. If crack sealing is required, care must be taken to choose the correct flexibility of injection resin (Section 11.9.4). Alternatively, a mastic-sealed formed joint may be created over the crack. Seek advice from the membrane manufacturer about the crack bridging qualities of their particular membrane systems, and ask to see test data to BS EN 1504-2 and their CE marking certificate, along with warranties etc.

Alternatively, where cracking is expected, proprietary movement joint systems should be provided, terminating the membrane either side of the joint (Fig. 11.2).

Concrete detailing should take account of any waterproof membrane system to be used, avoiding sharp corner edges and specifying using 45° fillets to upstands etc., as necessary. Upstands should not be coated to any appreciable height, as these may be subject to damage from vehicles and trolleys.

11.10.6.3 Concrete deck waterproofing (using a membrane)

A car park waterproof membrane should have the following properties:

- Capacity to bridge live structural cracks which open up after waterproofing and may be subject to rapid cyclic movement
- Chemical durability and compatibility with the joint materials with which it comes into contact
- Capacity to be bonded to concrete and/or capable of performing unbonded
- A surface that is skid- and slip-resistant and capable of resisting the abrasion and loadings from vehicles and trolleys
- Tolerance to being laid during local weather and/or to a damp concrete surface
- Resistance to direct solar exposure and maintaining its flexibility under freezing conditions (top decks of car parks only)

11.10.6.3.1 Liquid-applied polymer membranes (bonded)

The membrane element of a system for vertical upstands and horizontal surfaces commonly has a minimum dry film thickness of at least 1mm (often up to 5mm where reinforcing glass-fibre scrim reinforcement is used).

A ductile and resilient liquid-applied membrane system usually comprises at least four layers: primer, waterproof layer, and two coats of wearing surface incorporating non-slip aggregate. The first wearing coat should act as a matrix to hold the aggregate in place, while the second encapsulates the aggregate and provides a consistent appearance. Scrim reinforcement will usually be included within the waterproofing layer.

The final surface finish should be skid- and slip-resistant and ideally available in different light-stable colours to differentiate between parking bays and traffic aisles. It should be checked that the material used for line markings is compatible with the membrane system. Many systems use line-marking materials of the same generic type as the membrane.

The wearing surface must be designed to withstand abrasion and loadings from the volume of pedestrian and vehicular traffic expected. The system should also be capable of dealing with the variable abrasion conditions at turning areas, ramps, aisles, parking bays and kerb upstands. Suitable protection to edges of kerbs, etc., may be required if the membrane is not capable of resisting scuffing from vehicles.

As a minimum, the properties of the membrane should comply with Principle 1, Method 1.3 in BS EN 1504-2[135], with crack bridging requirements and skid-resistance selected to meet the likely exposure temperature and crack movement range at the car park site. Compliance with a relevant performance standard, that includes suitable durability test results and/or documented in-service performance, is therefore needed as the basis for material selection.

A liquid-applied membrane certified to comply with BS EN 1504-2 will carry a CE/UKCA mark verifying the product is suitable for the intended purpose as a protection system for concrete. In addition to meeting the listed mandatory requirements of BS EN 1504-2, the membrane is expected to meet the following criteria for UK and other Northern European conditions:

- Water vapour (S_d) should be Class III
- Skid-resistance for external use should be Class III
- Crack bridging classification:
  - Table 6 (of BS EN 1504-2): Static cracks – Class A5, bridging 2.5mm crack, opening at 0.5mm/minute, at -10°C
  - Table 7 (of BS EN 1504-2): Active cracks – Class B3.1, crack opening from 0.1–0.3mm width, 0.03Hz, 1,000 cycles, at -10°C

Note, that BS EN 1504-2 was not specifically intended for car park deck membranes, and is commonly used for coatings applied to non-trafficked surfaces such as walls and soffits of slabs. For the severe exposure conditions typically found on roof decks, the minimum performance requirements of EN 1504-2 may not be adequate. In some car parks, the crack widths may exceed the typical levels indicated, creating a risk of failure. In such circumstances, a reinforced membrane system may be required that is outside the scope of BS EN 1504-2.

Guidance relating to liquid-applied membrane application can be found in BS EN 1504-10[130]. The concrete surface should be finished to a smoothness suitable for the type of membrane to be applied, as set out by the manufacturer, but is typically a Basic or Ordinary finish (Table 11.5). As this type of membrane is fully bonded, it is vital that the surface is properly prepared by vacuum shot blasting or other similar means to remove all oil, grease, dust and laitance before laying; this operation is usually the responsibility of the specialist waterproofing contractor.

While waterproof membranes applied to the top deck and top ramps receive the most severe exposure conditions — including thermal movements that induce crack opening, and solar radiation that accelerates embrittlement — intermediate deck and ramp membranes receive less extreme exposure. The appearance of the car park can be greatly enhanced with intermediate deck coatings. Some of these are not full elastomeric membranes and often do not have the same crack-bridging capabilities as the spray-applied or thin-poured epoxy membranes. Such deck coatings would not be expected to be fully waterproof, unless specified as such. These materials offer limited protection against chloride ion ingress but enhance the environment and dynamics inside the car park. Other key properties, including adhesion to concrete and resistance to shearing under car tyre action, should not be compromised.

Brightly coloured membranes, coupled with decorative coatings (Section 11.10.6.4) can be highly effective in maximising illumination and reflectance, particularly in basement areas. Appropriate ventilation is needed when these are to be applied in confined spaces (Section 11.10.6.3.3).

11.10.6.3.2 Mastic asphalt

Traditional mastic asphalt waterproofing can provide a cost-effective solution, provided care is taken during preparation and application. The material is particularly sensitive to reflective cracking and so care is needed to seal passive cracks and fill depressions that could form points of failure. Guidance on mastic asphalt wearing courses is given in BS 8204-5[136].

Mastic asphalt should not be bonded directly to the concrete deck where it is exposed to direct sunlight. An underlay of sheathing felt should be used or, for partial separation, a felt or mat of woven glass fibre. This is important to prevent reflective cracking over any moving cracks in the concrete.

On ramps where the gradient is less than 1:10, mastic asphalt can be used but it must be bonded to the concrete. The concrete surface should be prepared by tamping or stippling to provide a key for the asphalt. Special care and detailing is needed at the intersection of ramps and floors and any cracks should be sealed with injection resin to prevent reflective cracks.

The life of the mastic asphalt depends on the formulation, with thin membranes having an expected life of up to 15 years, whereas 20 years is possible for some polymer asphalt systems[29].

Asphalt materials can yield under the combined effects of loading and temperature increase. They are not suitable as a founding layer for impact-resistant barriers, and consideration should be given to the use of plinths to raise any baseplates and fixings above the level of the waterproofing. When laid over insulating roof screeds subject to solar gain, asphalts may overheat and soften in service, leading to splitting and rutting.

Bituminous materials should not be used with polysulphides, since uncured polysulphide and bitumen are mutually soluble, leaving the cured material weak at the interface. Tar-based surface treatments are also unsuitable over membranes of synthetic rubber. All materials should be checked for compatibility and evidence of successful previous use.

11.10.6.3.3 Health and safety

Where membranes are to be applied in confined spaces, fumes can build up, and respiratory equipment or forced ventilation may be required. Reference should be made to the manufacturer's safe usage guidelines and, in the UK, to the Control of Substances Hazardous to Heath (COSHH)[137] legal requirements, so that works may be specified safely.

When considering membranes, fire resistance should be taken into account. CE/UKCA marking of liquid-applied membranes to BS EN 1504-2 includes testing to establish the Euroclass for fire resistance to EN 13501-1[138].

11.10.6.3.4 Quality Control

As a minimum, a liquid-applied membrane system should:

- Be UKCA-marked to comply with BS EN 1504-2 or hold an independent test certificate covering the intended use and relevant characteristics when applied to concrete (e.g., EOTA certification[139] or equivalent)
- Enable the manufacturer to demonstrate the successful application of the system on similar sites that have been in place for at least five years. This should be backed by independent performance-related documentation

11.10.6.3.5 Maintenance

Regular inspection is important to ensure that the waterproofing is fulfilling its requirements. Where necessary, maintenance/repair should be carried out strictly in accordance with the manufacturer's/installer's recommendations. The formulations of many thin-layer polymeric membranes are such that they are difficult to overcoat once aged, and specialist advice will be needed from the manufacturer. Repair of mastic asphalt surfaces is much more straightforward.

11.10.6.4 Decorative, protective and intumescent coatings

Good-quality concrete with adequate cover to reinforcement, complying with BS 8500-1[106] does not require any special paint finish for protection against standard exposure conditions. However, car parks are often painted to enhance their appearance and improve lighting levels. The use of bright colours, painted walls, decks and soffits can do much to obviate dark areas and reduce opportunities for crime. Guidance on painting concrete is given in BS 6150[134] and The Concrete Society's TR50[140].

Special consideration is required when painting external basement walls, depending on the water table. If the concrete is likely to be saturated, the paint system must have a low water vapour diffusion resistance if blistering and failure are to be prevented. This is a performance requirement for SPSs within the scope of BS EN 1504-2[135] (Principle 1, Method 1.3) and should be specified to be CE/UKCA marked in accordance with the standard.

11.10.6.4.1 Decorative coatings

Most decorative paint can be applied by brush, roller or spray. Good surface preparation is vital. All surfaces should be clean, dry and free from cracks or defective areas[130]. All cracks and defects should be repaired before treating the surface with a fungicidal wash, if this is required. When the concrete surface is porous, special primers may be needed before the final coat is applied.

Special anti-graffiti paint systems are available to protect vulnerable areas of the car park: these either form an impervious seal that can be cleaned with solvents, or are sacrificial and can be easily removed and replaced.

11.10.6.4.2 Protective coatings

These coatings have been specially formulated to resist acidic gases, chemical attack and water ingress and are within the scope of BS EN 1504-2. Surface preparation is similar to that for decorative painting, although where anti-carbonation coatings are required it may be necessary to fill blow holes and imperfections before painting, to provide a continuous film of paint.

11.10.6.4.3 Intumescent coatings

Intumescent coatings provide fire resistance to steel-framed structures. Consideration needs to be given to the moisture content and preparation of the steel for successful application. Advice should be sought on the suitability of coatings for the corrosive environment and, in particular, their resistance to abrasion. Accidental damage or vandalism could remove the coating and thus compromise its performance in a fire.

11.11 Temporary and demountable car parks

11.11.1 General

Temporary and demountable car parks are generally modular structures which comprise a series of standard components that can be dismantled and reused on multiple occasions. They typically comprise single storey structures for construction over an existing surface car park to provide additional parking. Generally, they are founded on spreader plates with no connection to the existing surfacing and have levelling jacks built into the columns to adjust for any variations in the surface levels and also to correct any movement that may occur as a result of settlement.

Temporary and demountable structures[141] are not limited to a single storey, and foundations may or may not be required, depending on the ability of the existing surface and substrate to support the structure without excessive settlement and the ability of the structure to accept the settlement without distress.

11.11.2 Performance criteria

To make temporary and demountable car parks economically viable, the design service life is often shorter than those of permanent car parks. However, the quality and the design life of some demountable structures can be of an equivalent standard. Whatever the design service life, the temporary structure should be designed to the same loadings and conditions as a permanent car park, including edge protection and measures to prevent progressive collapse from accidental and other loading.

Skeletal cladding is a popular choice for this type of structure, and consideration should be given to the additional loading from icing during winter, especially in areas subject to driving rain and snow.

The structural design and choice of materials and finishes should be suitable for the intended purpose. Sharp edges, trip hazards, poor slip resistance and poorly located bracings which impose a risk to personal injury, should be avoided.

Integrated drainage systems are necessary for temporary structures and the decks should be laid to falls to prevent ponding. Surface water from the elevated deck can discharge onto the existing surfacing, provided it is properly managed to avoid any ponding and is collected by the below-ground drainage system.

All decks and joints between the components should be properly sealed to prevent uncontrolled leakage to the floor below. Where waterproof membranes or deck coatings are applied the slip-resistance should be of low slip potential and have a pendulum test value of 36 or greater[142].

Fire protection and means of escape provisions should be identical to those for permanent structures where the structure is more than 1-storey. For single storey open deck car parks the travel distances can be enhanced (Chapter 14).

11.11.3 Inspection and maintenance

Temporary and demountable car parks must be inspected and checked at regular intervals by a competent person to ensure that they have not been subject to accidental damage, and all bolts are in place and tightened to the correct torque. Levels and vertical measurements should be taken at column positions to check that the structure is performing within its design settlement criteria, and any necessary adjustments should be made to prevent over-stressing of the structural members and connections.

12 Guarding and containment

12.1 Introduction

Car parks are dynamic buildings where people and vehicles move horizontally and vertically; and measures are required to ensure that this can happen safely. As it is difficult in a shared space to separate pedestrians from moving vehicles, signs are displayed to limit the maximum speed to 5mph (8km/hr). This is to minimise the risk of serious injury should a low-speed impact occur. Children are the most vulnerable group and can be unpredictable and run between parked vehicles and into the drive aisles. They can even attempt to climb the guarding that is intended to keep them safe. Those looking for a place to self-harm also need special protection (Chapter 16).

While car parks of three or more storeys without any measures to break a fall present the greatest risk, an assessment should be carried out for all car parks to identify the foreseeable risks and to configure any mitigation measures required. A high fence around the perimeter of the top deck is often seen as an effective solution, but the risk of moving to the level below must not be overlooked. High fences do not create a pleasing aesthetic, and improving the environment through good design is often seen as a better solution. Adding a roof over the top deck with a mansard or perimeter screens, good quality lighting, good surveillance (rather than denying access to the roof) and playing soothing music will help to create a calming environment. Having appropriate signage and a helpline is also beneficial to those in need of assistance.

Car parks are alternative environments for skateboarders especially if the decks are inclined, and they can be at risk of flipping over the barrier especially if they approach it at speed. Proportionate containment measures are therefore required to ensure the safety of all.

Containment is also fundamentally required to ensure vehicles of all types do not leave the edge of the car park at height (Section 12.2). These containment barriers are very different in nature to pedestrian guarding (Section 12.3).

12.2 Vehicle restraint barriers

12.2.1 Types of edge protection

There are many different types of edge protection but essentially, they fall into the following groups:

- **Rigid barriers** i.e., those that are non-yielding under impact e.g., reinforced concrete upstands that cantilever from the deck (Figure 12.1) or steel hollow sections or Armco® barrier rails on rigid posts bolted to the deck (Figure 12.2).
- **Flexible barriers** i.e., those that yield under impact to absorb some of the kinetic energy and reduce the effective load on the vehicle e.g., semi-rigid barriers comprising steel or polymer-modified plastic barriers on flexible posts, or steel sections that span between building columns (Figure 12.3) or flexible spring steel posts or posts with rubber inserts that deflect and absorb some of the energy. Large energy-absorbing systems tend to comprise tensioned steel cables that span between building columns, or wire mesh systems that fix to the building fabric (Figure 12.4).
- **Proprietary systems** where the barrier is integral with the pedestrian guarding or where the barrier is integral within the building cladding (Figure 12.5).

It is possible to design rigid barriers by calculation as the loads can be easily defined, but flexible barriers are energy-absorbing systems that have been developed and certified after extensive compliance testing.

Containment to restrain an errant vehicle is required in the following situations:

- To edges of decks and ramps where there is a change in level
- To protect fragile cladding from displacement
- To protect party and fragile internal walls around stair cores, lift shafts, offices etc. from collapse
- To protect vulnerable columns from damage that could trigger a disproportionate collapse
- To protect sensitive equipment
- To protect those in the immediate vicinity of the car park

Figure 12.1: Reinforced concrete cantilever upstand restraint (note cracking from vehicle impact)

Figure 12.2: Rigid post cantilevered from deck after impact on rail

Figure 12.3: Hollow section vehicle restraint spanning horizontally between columns

Figure 12.4: Tensioned wire restraint barrier

Figure 12.5: Vehicle restraint barrier integral with cladding

12.2.2 Required performance

While the design loading is given in BS EN 1991-1-1[74] and BS EN 1991-1-7[78] (including the relevant UK National Annex to Eurocodes[143]), and Approved Document K of the Building Regulations[144] and is based upon absorbing the kinetic energy from the moving vehicle, the increase in weight of hybrid and battery electric vehicles has brought into question the validity of the design criteria to cater for the current range of vehicles on UK roads. Reference should also be made to the UK Government report: *Edge protection in multi-storey car parks – Design specification and compliance testing*[145] for modification to the dimensional design criteria in the Eurocodes and the Building Regulations.

Vehicle restraint barriers are required to resist a 2,000kg vehicle travelling at 4.5m/sec (10mph) without excessive deflection. They can deform permanently requiring replacement, but they must not fail. A clearance gap sufficient to allow the barrier to deflect is required between the back of the barrier and any cladding to ensure that the cladding cannot be dislodged from the car park especially if there is pedestrian access below, and therefore a risk of personal injury from a falling cladding panel. The deflection is limited to 600mm – to ensure that the vehicle wheels cannot travel beyond the edge of the deck, and that pedestrians cannot fall through the gap between the deflected barrier and the exposed edge of the deck.

It is essential to ensure that the energy imparted by an errant vehicle to a barrier or column does not result in any large-scale damage or disproportionate collapse of the structure. Foundations to columns and structural members are required to resist the forces from vehicle restraint barriers or systems connected to them.

The height and effective width of the barrier system should ensure that vehicles cannot travel under or over the barrier. When tensioned cables are used, there should be a means of linking the cables to prevent them from moving apart under impact (Fig. 12.4). The effective height of the centre of the barrier rail or barrier system should be 445mm above deck level which is considered to be the front bumper height of the standard design vehicle (Section 5.2.1). Where a concrete upstand provides vehicle restraint, the top should be a minimum of 600mm above deck level.

Vehicle restraint barriers must not be an invitation to climb especially for young children and should not allow pedestrians to roll between the bottom of the barrier and the deck (Section 12.3.1). The barrier system should be configured to prevent it from being used as a step (creating a risk of persons falling from the car park). The minimum height of the top of the barrier should be 550mm above deck level to minimise the potential step-up. Where the step is less than 550mm, the top of any parapet or handrail that can be reached from the step should be increased to a minimum of 1,100mm above the level of the step. If infill mesh is used between the handrail and the vehicle restraint barrier to prevent pedestrians from being able to achieve a foothold, the height of the handrail can be 1,100mm above deck level.

At split-levels where a substantial part of the vehicle could pass through the gap between the levels, a vehicle restraint system should be provided (the deflection of which should be limited to ensure that pedestrians on the level below are not injured). In this position, wire mesh or similar combined pedestrian and vehicle restraint barriers are often provided.

Guarding and means of containment should be configured to not obscure visibility on ramps and where clear lines of sight are essential.

The horizontal force *F* (kN) normal to a barrier or column is given by:

$$F = \frac{0.5mv^2}{\partial a - \partial b}$$

Where:

m = gross mass of vehicle (kg)

v = velocity of vehicle normal to barrier or column (m/sec)

∂a = deformation of vehicle (mm)

∂b = horizontal deflection of barrier (mm)

Where the car park is designed on the basis that the gross mass of the vehicle (vehicle plus passengers and luggage) will not exceed 3,000kg the following values are used to determine *F* which can be taken over a 1,500mm horizontal length of barrier:

m = 2,000kg[†]

v = 4.5m/s

∂c = 100mm unless better evidence is available

For a rigid barrier or column, for which ∂b may be taken as zero, *F* appropriate to vehicles up to 3,000kg gross mass is taken as 200kN.

Where the car park has been designed for vehicles with a gross mass exceeding 3,000kg, the following values are used to determine *F*:

m = actual mass of vehicle for which the car park is designed (kg)

v = 4.5m/s

∂c = 100mm unless better evidence is available

F is applied at bumper height which is taken as 375mm above deck for vehicles with gross mass up to 3,000kg. However, the UK Government report: *Edge protection in multi-storey car parks — assessment method for installed restraint systems*[146] concluded that the average front bumper height is 445mm above deck level. This figure is now considered to be the standard bumper height and should be used in the design, testing and certification of barrier systems. Although the rear bumpers of many vehicles can be higher than the front bumpers, 445mm is used as the impact height above deck level.

[†] The average mass of vehicles in previous guidance was considered to be 1,500kg, but as vehicles are continuing to grow in size and their means of propulsion is changing from internal combustion engines to batteries, which are significantly heavier, the increase in mass is approx. 500kg. A mass of 2,000kg is therefore considered to be more representative of the average vehicle than the extreme value of 3,000kg.

It can be difficult to design the fixings to the decks for rigid barriers with non-yielding posts as the forces are too large for most holding down bolts, particularly when the impact is directly at a post position. It can therefore be assumed that the impacted post fails and the barrier rail acts in catenary, transferring the load into the adjacent two posts, thus halving the loads on the posts.

12.2.3 Barriers at the ends of long straights

When the vehicle approach length to a barrier (i.e., at the ends of long drive aisles, parking ramps or cul de sacs) exceeds 20m in a straight line, the barriers are required be designed for twice *F* acting at 445mm above deck level. Where it is considered that vehicles are likely to significantly exceed the speed limit within the car park, speed retarders or other measures should be introduced to reduce vehicle speed (Section 5.4.3).

12.2.4 Barriers in close proximity to vehicle ramps

Barriers within 5m of a vehicle ramp (excluding parking ramps) that could be impacted while approaching or leaving the ramp, and barriers to the sides of vehicle ramps, should be designed to withstand half of *F* acting at a height of 610mm above the running surface.

12.2.5 Barriers opposite the ends of long vehicle ramps

Barriers at the ends of straight ramps which are more than 20m long with a gradient in excess of 1:20 and intended for downward travel, are required to withstand twice *F* acting at a height of 445mm above the running surface of the ramp.

12.2.6 Fixings for vehicle restraint barriers

Fixing bolts which the barrier relies on for attachment to the structure must not fail or pull out (Figure 12.6) and should be configured to minimise the loading on the column to prevent permanent damage or progressive collapse of the structure. However, where locating bolts are used to fix the barrier to the car park columns, they may be designed to fail in order to minimise damage to the columns, provided the barrier remains contained in a fail-safe configuration e.g., between column flanges.

Cantilevered barriers attached directly to the concrete deck should have fixings that are rigidly anchored into the concrete with a factor of 1.5 applied to the design forces in the bolts, to ensure they do not fail and can be reused if the barrier needs to be replaced. The capacity of the deck to resist the localised bolt forces without failure needs be confirmed, and in thin decks, through bolts are often used in conjunction with plate washers on the underside of the deck.

Figure 12.6: Flexible post cantilevered from deck (note bolt failure after load test)

Other types of proprietary anchors may be suitable, but their ability to remain anchored into the slab under successive load applications must be demonstrated. This is to prevent minor impacts reducing the fixing capacity without that reduction being apparent before a significant impact occurs. Suitability of fixings can be confirmed by tests that repeat the predicted combination of loads four times to determine their actual factor of safety.

Setting barrier posts on plinths will enhance durability especially if the decks fall towards the barrier and drain in this area. Holes drilled for fixings should be positioned to avoid reinforcement. For through-slab holes, diamond drilling causes less soffit breakout than percussion drilling. While sealing the bolt into its hole will help prevent water ingress, stainless steel components offer the best protection against corrosion.

12.2.7 Compliance testing and certification

The satisfactory performance of all barrier systems should be capable of being demonstrated by calculations or by compliance testing. *In situ* static load testing will usually give a conservative assessment of the load carrying capacity and is more suitable for rigid or semi rigid barriers. This is because the apparent strength of materials increases with the rate of strain.

Where calculations are not possible owing to the barrier being a proprietary or large energy-absorbing system, dynamic testing is required to determine the performance, but owing to the size and operation of the test rig it is not suitable for testing within the car park. Recommendations for dynamic testing are given in *Edge protection in multi-storey car parks — assessment method for installed restraint systems in multi-storey car parks*[146]. The testing and certification of vehicle restraint barriers should be undertaken by an accredited testing house.

Barrier fixing bolts may also require pull-out tests to be undertaken to demonstrate their capability to resist the forces imposed, especially when anchored into thin or voided decks (Section 12.2.6).

12.3 Pedestrian guarding

12.3.1 Purpose of guarding

Pedestrian guarding is required to protect people in, or in the immediate vicinity of, the building (lawfully or otherwise) from the risk of personal injury. The type of guarding and choice of materials should be suitable for the purpose intended and should prevent people from falling, rolling, sliding, or slipping through or under the barrier. As young children are considered to be vulnerable, the guarding should not be an invitation to climb or easily climbable. Horizontal rails should therefore be avoided unless an infill mesh is provided to minimise the opportunity of gaining a foothold.

12.3.2 Required performance

In accordance with Part K of the Building Regulations[144] pedestrian guarding is required at changes in level which are greater than 380mm. However, in public buildings where users may be less familiar with their surroundings (which is considered to be the case for car parks) and in situations where users may have mobility or sight issues, consideration should be given to guarding all changes in level e.g., to the edges of decks, ramps, split-levels and stairs, landings and in front of openings to which people could gain access. Deterrents should be incorporated to avoid the risk of jumping into any voids, especially within stair cores. The height should be a minimum of 900mm to stairs and 1,100mm to landings, decks and ramps. Consideration should be given to increasing the height of pedestrian guarding to 1,500mm above deck level where the car park is adjacent to an operational railway.

Where there is a risk of jump-related suicide, refer to Chapter 16.

Where tall or full height mesh screens are used, consideration needs to be given to ensure that the natural ventilation of the car park and the accessibility to fight fires and safely rescue occupants is not adversely affected. Should this be identified as a potential risk it may be necessary to incorporate mitigating measures which would need to be designed holistically as part of the fire safety strategy.

Where pedestrian guarding is used in conjunction with vehicle restraint barriers it is essential to ensure that a foothold cannot be gained to step up on the restraint barrier. The pedestrian guarding needs to be configured to ensure that either the potential for a foothold is minimised by obscuring the vehicle restraint barrier with an infill mesh or increasing the height of the guarding to a minimum of 1,100mm above the potential step-up (Section 12.2.2).

The maximum gap between vertical rails in pedestrian guarding should not exceed 100mm.

The minimum service loading is given in BS EN 1991-1-1[74] (including the relevant UK National Annex[75]) and Approved Document K of the Building Regulations and is summarised as:

Horizontal uniformly distributed line load: 1.5kN/m

Uniformly distributed load applied to infill: 1.5kN/m^2

Point load applied to infill: 1.5kN

Standard loadings may also need to be increased where there is a likelihood of crowd loading from people using the car park as a viewing point.

Guidance on the construction and configuration of guarding is given in BS 6180[147] which states that the total horizontal displacement of the guarding at any point from its unloaded position should not exceed 25mm. The displacement of any point of the infill should not exceed $L/80$ or 25mm (whichever is the lesser) where L is the clear span of the panel between supports. These limitations on deflection are also intended to apply to combined vehicle and pedestrian guarding while subject to pedestrian loading only, but in the event of a vehicle impact the limitations on deflection are as specified in section 12.2.2.

12.3.3 Holding down bolts

Cantilevered barriers attached directly to the concrete deck should have fixings that are rigidly anchored into the concrete, and a factor of 1.5 is usually applied to the design forces in the bolts to ensure they do not fail and can be reused if the barrier needs to be replaced.

While in service, the holding down bolts will undergo many cycles of loading and unloading and therefore the choice of bolts should ensure that they do not work loose.

13 Integration of building services (detailed design)

This chapter builds on information in Chapter 6 to provide more detailed guidance on the design of the key building services elements, as well as providing useful reference for the building services consultant in the context of integrating these systems within the overall car park design.

13.1 Mechanical services

13.1.1 Heating

Car parks are typically considered as unoccupied and therefore have limited requirements for heating. Frost protection requirements may involve the use of ramp heating systems — although it is generally preferable to avoid places where water can accumulate on roadway surfaces rather than use these systems, which will increase energy use in addition to running costs.

Staff areas and ancillary spaces (if considered occupied) should be treated as usual and the appropriate guidelines on suitable environmental conditions should be followed. Note, that staff may occupy spaces such as ticket booths for extended periods, carrying out largely sedentary tasks, and the environmental criteria should be chosen accordingly[148].

13.1.2 Ventilation

A natural ventilation solution should always be prioritised, minimising the use of any mechanical systems, and where possible this should be factored into early considerations. Influencing the design by reducing the use of basement car parks supports the use of a natural ventilation solution and supports more sustainable projects with the reduction of energy intensive basement building.

Car park ventilation has been required in the UK for a number of years, to monitor and control the build-up of vehicle exhaust fumes. Traditionally, where mechanical ventilation has been required, it was common for it to be a ducted exhaust system with multiple inlets throughout the space. More recently there has been a switch to impulse/jet fan ventilation combined with a single point of extract. Smoke ventilation may also be required in the case of a fire to ensure safe evacuation and smoke clearance.

The requirements for car park ventilation are derived from Building Regulations' Approved Document B[66] and Approved Document F[149] in England which specify protection against fire and general ventilation respectively. The day-to-day ventilation criteria relates to the maximum level of carbon monoxide (CO) permissible in a time frame.

The risks associated with a vehicle fire can be very significant from both heat and smoke. Car park systems designed in accordance with current building regulations are designed to aid the fire service attending an incident. The smoke system is designed for smoke clearance only, and further steps can be taken to engineer a solution for smoke control.

BS 7346-7[150] covers the full range of ventilation options; from full natural ventilation to comprehensive smoke control, making provision for life safety.

The main types of ventilation system likely to be adopted for car parks are:

- **Natural ventilation**
  Natural ventilation may be used in car parks provided 5% of the floor area acts as a free area of permanent wall openings to provide cross-ventilation at each level — to fulfil both general and smoke clearance requirements[66, 149]. At least 50% of the free area (i.e., 2.5% of total floor area) must be arranged equally across opposing sides to assume that ventilation is well distributed and effective. However, care must be taken to ensure that no stagnant areas remain that may require further openings or local mechanical ventilation to resolve. Reversible jet fans can be supplemented to enhance the ventilation performance and reduce stagnant area

- **Mechanically-assisted natural ventilation**
  If the natural ventilation requirement of 5% of the floor area as a free area of permanent wall openings cannot be met across the car park, a hybrid solution can be implemented. Permanent natural ventilation with an aggregate equivalent area of at least 2.5% of the floor area can be combined with a mechanical system capable of at least three air changes per hour (ACH). Exits and ramps should be equipped with a local ventilation rate of at least 10 ACH. A unidirectional jet fan is required for achieving this design principle

- **Ducted exhaust systems**
  These systems utilise ductwork distributed throughout the car park to ensure all areas are sufficiently ventilated. Ductwork sizes can often limit the available clear height and can block out light as well as the path of security cameras. The lengths of ductwork required to prevent stagnant air reduces efficiency of the fan. For these reasons, ducted solutions are better suited to smaller car parks. Figure 13.1 shows the layout of a typical ducted solution

Figure 13.1a: Typical mechanical ventilation using a ducted smoke clearance system: plan view

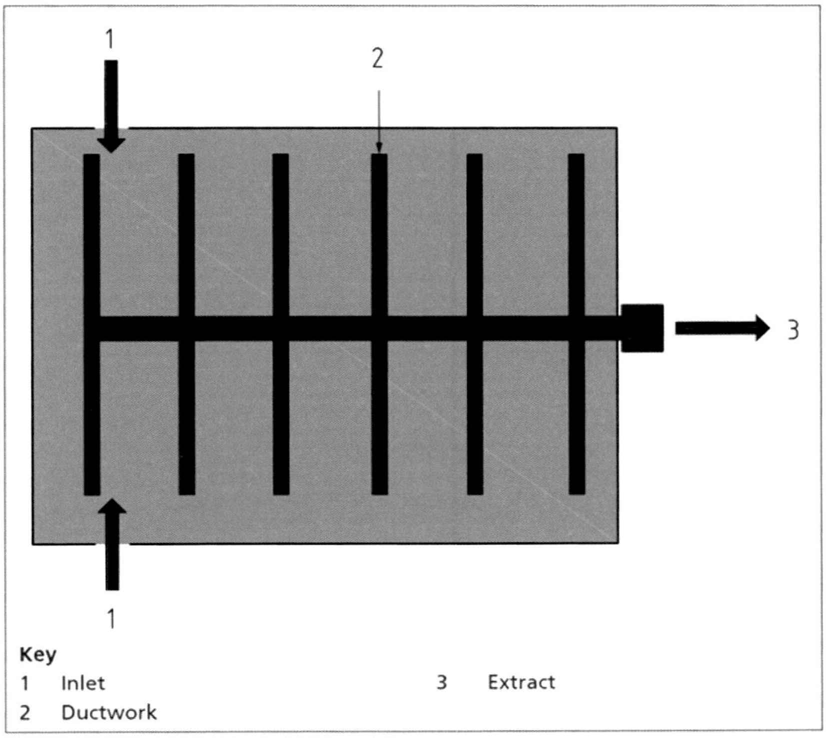

Figure 13.1b: Typical mechanical ventilation using ducted smoke clearance system: section view

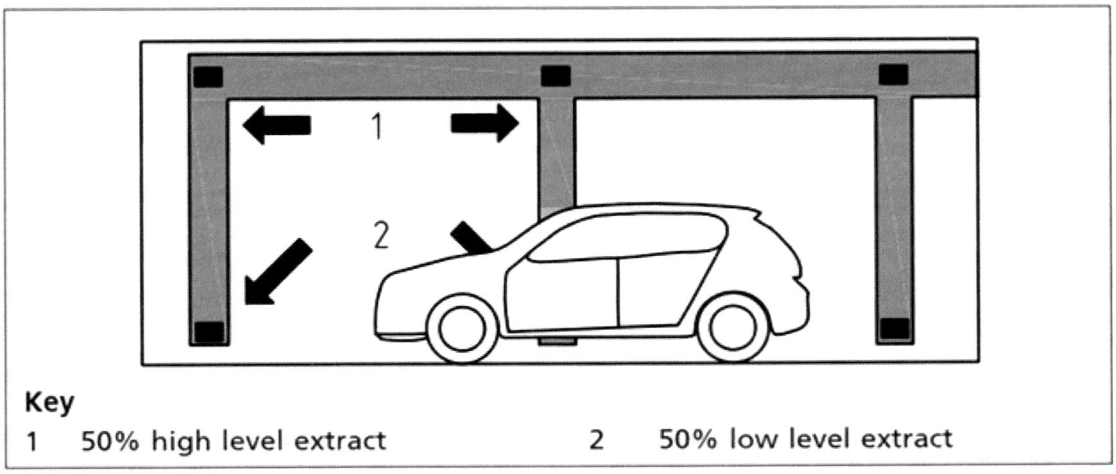

- **Impulse/jet fan systems**
  As an alternative to traditional ducted systems, car parks can be ventilated by an extract fan typically using a single extract shaft. This is in conjunction with impulse/jet fans at high level within the car park (Figure 13.2). A jet fan is made up of either a centrifugal fan intaking air from below and discharged at the side, or an axial fan. Jet fans are positioned so that their flow is discharged at an angle. They typically move a low flow of air but as it expels from the outlet at a high velocity, it drives air movement within the space (Figure 13.3)

Figure 13.2: Typical mechanical ventilation using impulse smoke clearance system

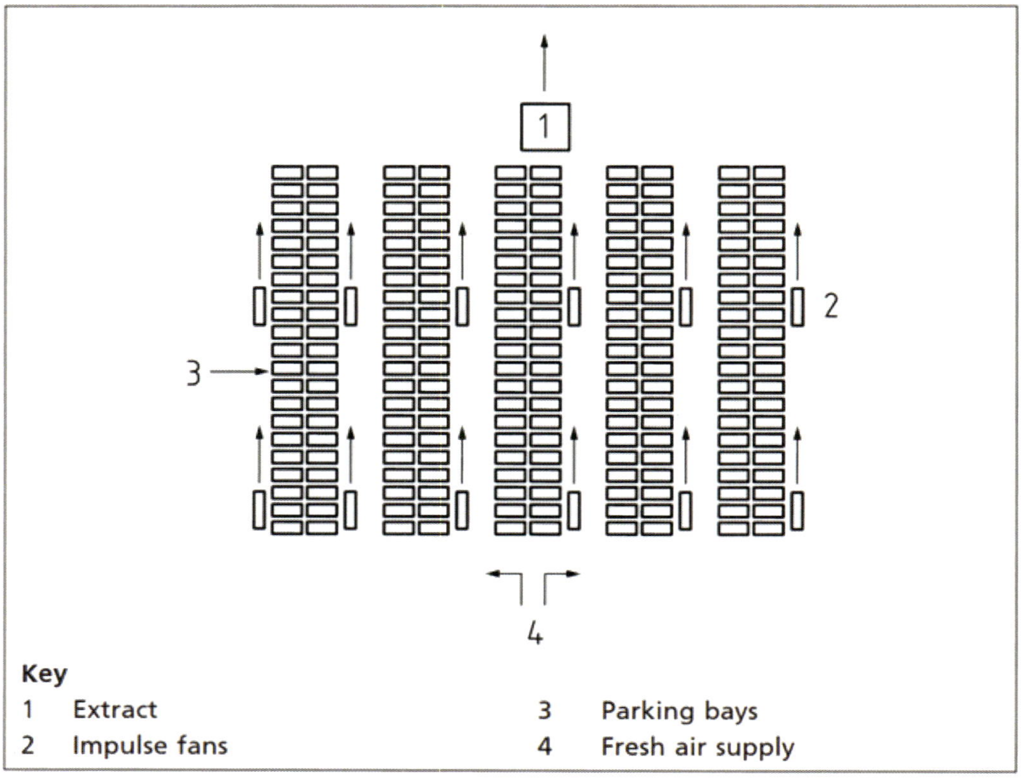

Figure 13.3: Jet fan system

13.2 Electrical services

13.2.1 Lighting

Although some car parks may benefit from being partially day lit, it will usually be necessary to provide lighting for continuous use, at least in some areas. It may be possible to make use of daylight for part of the year. The lighting controls and luminaires can be designed to allow advantage to be taken of available daylight, thus reducing operational costs.

Consideration should be given to control luminaires by the use of movement detectors with run-on timers. This system would typically control half of the luminaires with others being constantly illuminated. Typically, movement detectors are ultrasonic. This system cannot be considered if discharge lighting is specified, but LED lighting is likely to be considered in the majority of instances.

Lighting should be sufficient to assist the safe movement of vehicles and pedestrians and enable staff to carry out their duties. It should also be of a standard that will help reduce vandalism and other criminal activity. Luminaires with steel bodies, and polycarbonate diffusers with non-standard screws should be considered. The quality of the lighting will have a strong effect on the attitudes of users and will affect both the degree of vandalism and the general care taken with the building (Fig. 4.3). Lighting has a central role in the Park Mark® scheme[17].

As well as being vandal-resistant, lighting units need to be resistant to water ingress from the use of pressure sprayers to clean/wash down the car park. It is recommended that the minimum IP rating considered should be IP X5 to allow for standard cleaning regimes.

The nature of car park structures usually inhibits the scope of the lighting designer, but nevertheless an early association with the architect and structural engineer may produce, within a reasonable budget, a better-considered, visually attractive, easily maintained lighting scheme.

A particular challenge is created in car parks by the low headroom and structural form that make it difficult to provide the reasonably even lighting that would be expected in commercial buildings. Research in the UK on acceptable lighting quality for multi-storey car parks has led to a marked increase in recommended uniformity (ratio of minimum lighting level to average lighting level in a specified area) to 0.4, against a previous recommendation of 0.04. But a uniformity value of 0.4 should be seen as a minimum requirement and the designer should seek to improve on this wherever possible, with target levels of up to 0.7/0.8 in parking bays.

In addition to influencing a building's commercial success, high-quality lighting can be a major factor in obtaining acceptance of a multi-storey car park near high-quality or historic buildings. At least one city (Lyon, France) has made high-quality lighting of car parks mandatory as part of a campaign to improve the night-time visual environment.

A multi-storey car park can form a major part of the night-time scene and care is needed to make the structure aesthetically acceptable by means of appropriate lighting, and to ensure that light pollution and obtrusive light are kept to a minimum. Avoiding lighting pollution can be particularly difficult on open top decks.

Glare and colour rendering should also be considered in the selection of luminaires; both in relation to identifying hazards for drivers and in the context of CCTV system design. Colour recognition and the overall efficiency of lighting systems can be improved by painting walls, floors and ceilings in light colours and ensuring surface reflectance levels are high. Advise should be sought from the car park designer in determining appropriate reflectance levels to use within lighting calculations.

Many national lighting codes now require the designer to make assumptions during the design procedure on the maintenance regime to be followed. They are based on designing for a minimum maintained illuminance[151]. These assumptions are fed into the calculation procedure. To put it simply, if maintenance standards are expected to be high, less allowance has to be made in the design for dirty luminaires and lamps beyond their design service life. It is therefore essential that the client agrees on the maintenance regime before design begins, since the decision has major commercial implications for the car park operator. This maintenance regime should be contained within the operation and maintenance manual.

Many design guides for lighting exist which provide guidance and recommendations on the appropriate lighting levels, uniformity levels and design criteria for lighting installations in car parks[14, 15, 151, 152].

Particular attention should be given to achieving appropriate lighting levels within different areas of the car park (e.g., vehicle entry/exit points, ramps, changes of direction, cashier/ticket machine locations, parking and accessible parking bays, pedestrian areas etc.) linked to the specific activities at that location, and due consideration given to contrast levels with daylight.

The European Standard Parking Award Off-Street[16] provides a good point of reference for the lighting in such areas, as this recognised award scheme includes lighting criteria (average illuminance levels for various functional areas) as part of their checklist to obtain credits towards an overall award. Although credits are based on measured illuminance levels from completed car parks, it is nonetheless imperative to appreciate these at the design stage to ensure targets are embedded into the detailed design process. ESPA defines three levels of lighting, with decreasing credits awarded accordingly:

- An ideal level
- A lower range which is considered acceptable
- A level below which lighting would be deemed inadequate

Table 13.1 presents a summary of recommended illuminance levels for the most common areas within multi-storey and underground car parks.

In addition to typical lighting installations, car parks and their ancillary areas in many countries will be required to meet mandatory emergency lighting requirements. In the UK and Europe BS EN 1838[153] and BS 5266[154] apply.

Emergency lighting should be in the form of illuminated emergency legend panels and various emergency luminaires. System types vary from central battery to self-contained systems. Consideration on system selection should be given to ongoing system servicing and maintenance, balanced against the effect of carrying out this maintenance within the car park.

Table 13.1: Recommended illuminance levels

	Illuminance, average lux	Minimum uniformity	Comment
Parking bays (where open to public)	150	0.4	Can lower to 75 lux where non-public. Uniformity levels of 0.7/0.8 should be targeted
Access lanes, driveways	75	0.4	
Ramps, corners, intersections	150	0.4	
Entrance/exit zones (vehicular)	night 75 day 300	0.4 0.4	Reference BS EN 12464-1[14] Reference BS EN 12464-1[14]
Pay machine/cashier/barrier area	300	0.6	Care should be taken to achieve high vertical illuminances, with consideration of tasks
Pedestrian area, stair, lift	100	0.4	Higher within 1m of lift entrance (200 lux)
Roof level parking in rural zones E1 and E3	15	0.3	Reference BS 5489-1[15]
Roof level in urban zones E3 and E4	30	0.3	Reference BS 5489-1[15]

13.2.2 Power

While all car parks have a basic need for an electrical infrastructure, the extent of the electrical power provision is currently being driven significantly by the increasing demands of electric vehicle charging. Depending on the quantity and type of chargers provided, it is likely that peak electrical demands of EV charging are significantly higher than the rest of the car park's electrical demands, and therefore early consideration of requirements will be essential.

Understanding the maximum demand load of the facility will have significant influence on the discussions with utility suppliers on the type and size of incoming service (e.g., LV/HV) and therefore the space requirements of the infrastructure to serve the car park (e.g., substations, distribution equipment).

While considering maximum demand, opportunities for incorporating photovoltaics and battery storage technologies should be explored, as these could provide benefit in reducing peak demands and increasing operational energy efficiencies. This is particularly relevant given the increasing nature of electrical demand and load profiles in relation to EV charging. As part of assessing load requirements, it would be prudent to consider allowances for future demand changes, and to build in appropriate spare capacities within electrical infrastructure. This will minimise the risk of significant design change to accommodate changing technology (for example, future incorporation of additional high powered EV chargers).

There will be a requirement for most car parks to provide power to the service lighting, lifts, ventilation and other building systems required for the everyday function of the car park. In designing the electrical distribution, the need to meter different building uses should be considered, so that appropriate loads can be supplied by separate distribution boards (e.g., dedicated distribution for lighting, separate from ventilation plant). Care should also be taken in design to situate distribution boards and electrical panels away from vehicle and public areas to avoid increased risk of damage or vandalism.

Containment systems for cabling should consider robustness of design, both in terms of location and material selection, to minimise risk of damage due to vandalism, or unintended damage by vehicles.

Where life safety or firefighting supplies are required (most commonly for smoke control systems, fire suppression systems and firefighting or evacuation lifts) a secondary power supply will be required. BSRIA's BG 70/2021[155] provides information on designing power supplies for such systems, but most commonly this will involve the provision of a standby generator.

13.2.3 EV charging facilities

The EV industry is developing at significant speed. Battery technologists, charger technologists and car manufacturers are all working to develop their respective offerings and it is likely that as the technology and future trends develop, the infrastructure required to support them will also develop.

Design considerations around this subject are vast and cross multiple design disciplines notably including spatial layouts and fire protection provisions as well as building services. Key building services considerations are:

- **Primary distribution and load demand**
  As described in Section 13.2.2, early consideration of the power demands to be allocated to EV charging is key to determining the overall facility maximum demand load. Power distribution and infrastructure to support EV charging should be designed to deliver as a minimum, the overall capacity agreed as part of the load demand. In many cases, it is prudent to consider increased capacity within the distribution to cater for potential increasing future demands. As this will be part of the fixed distribution, providing a degree of flexibility for future change in the final circuits/charger connection points, may provide some ability to upgrade without significant re-cabling works. Providing a dedicated distribution for EV charging is advisable — both to segregate building uses and energy loads for metering, and to minimise the impact of any fault or maintenance on the remaining distribution systems

- **Charger connection points**
  These are the electrical points that provide connection to the EVs. They will be located adjacent to the car parking bays to facilitate connection to the vehicle. There are currently various products available providing a mixture of AC or DC charging. Some are specific to car manufacturers (e.g., Tesla) and some provide universal connection points that suit several vehicle manufacturers. The maximum capacities of these connection points will need to be matched to the final circuit allocations and electrical infrastructure design. Consideration for the location and planning of these points is key to ensure cable routes can be accommodated. Where pedestal type chargers are used it is likely that cable routes may need to be cast in the ground, so coordination with the structural engineer will be required.

  There is significant benefit in grouping charging points in a common location within the car park but consideration must be given to the loading on the structure, as battery-powered vehicles are heavier than their fossil-fuelled equivalents (Section 11.1.2). This will assist in minimising the electrical distribution to supply these points, and so early engagement with the car park designer to identify quantity and location of vehicle charging points is advised. Given the speed of development in the EV charging market, for larger projects with longer construction periods, it may be beneficial to defer the selection of charger connection points until later in the delivery programme to ensure the most up-to-date products can be selected

- **Smart charging technology**
  Many of the products currently available in the market permit the use of 'smart charging'. This is essentially control technology that enables vehicles to charge at varying capacities, depending on energy availability. Electric vehicle batteries can be charged in a smart way to prevent peak loads on the building infrastructure. This can be based on energy demand and available capacity on a local level, and can determine when, and at which capacity, the vehicle will be charged. Smart charging is likely to be required to ensure the overall capacities allocated to EV charging are not exceeded

13.2.4 Fire alarm systems

In the UK the installation of fire alarm systems is subject to the requirements of BS 5839[156]. The system design will vary depending on building design, usage and occupancy and the associated risks. Standard buildings may have straightforward fire alarm system solutions, but regardless, early alignment with the building fire strategy is key to ensuring the appropriate provision, and allowing for interfaces with other building systems (e.g., fire suppression, lift installations, security/access control and ventilation/smoke ventilation systems).

Key issues to consider in respect of fire alarm system design for car parks include:

- **Cabling arrangements.** The selection of cable types and their associated containment systems with particular attention to vandalism. Many car parks will be publicly accessible and with relatively low slab-to-slab heights, keeping cabling contained within enclosed trunking systems is often the most appropriate way to minimise risks. Where vandalism is not such a key consideration, cables may be clipped direct or installed on cable tray or basket systems. Fixings for these cabling systems should at least match the fire performance standard required of the cables themselves — this is discussed in BS 8519[157] and BS 5839
- **Appropriate selection of fire detection devices.** Within car parks the presence of car park exhaust fumes increases the risk of false alarms where smoke detectors are used, so it is appropriate to consider the use of heat detectors, or beam/optical detection as alternatives
- **Interfaces with other building systems.** Depending on the complexity of the building this may vary significantly, but typically fire alarm systems will interface with other systems such as:
  - Fire suppression
  - Lift installations
  - Security/access control
  - Ventilation and smoke ventilation/clearance
  - Motorised fire and smoke dampers
  - EV charging, and/or battery storage
  - Life safety or standby generator
  - Emergency/anti-panic lighting
- **Development of a cause and effect strategy aligned to the fire strategy.** This is particularly relevant for more complex buildings which incorporate car parks, because the greater the number of system-to-system interfaces, the greater the need for clarity around their cause and effect

13.2.5 Lightning protection/earthing

Although not usually the tallest buildings, multi-storey car parks should be equipped with appropriate lightning protection (particularly where the top deck and metal balustrading is exposed), subject to the completion of an initial risk assessment. Guidance on design is contained in a European code of practice which also contains information on protecting IT equipment against lightning strikes[158].

Lightning protection can take various forms but will usually consist of a network of copper tapes linking air terminations to buried copper earthing rods. It may be possible to make use of steelwork within the concrete structure as an alternative to copper tapes, but great care is needed to ensure that all relevant joints between reinforcing rods are bonded and of low impedance. It will in any case probably be necessary to bond the lightning protection system to the reinforcement. Bond is also required to other metalwork within the 'separation distance' of the lightning protection network. This will always include the utility services.

All earth bonding should be undertaken in line with the current edition of the IET Wiring Regulations[159] or local/national wiring regulations. Earth bonding should also be applied to a steel frame structure and ductwork systems.

13.3 Public health services (drainage)

Well designed and properly maintained drainage is essential to ensure that all water deposited on the exposed surfaces is rapidly discharged through an effective system. To achieve this, all drainage outlets should be recessed below the finished top surface of the concrete to ensure that no ponding occurs. Standing water is detrimental to both the structure and the operation of the car park. In the UK, drainage is designed to BS EN 752[101] and BS EN 12056[102, 103]; local standards will apply in other countries. Specific considerations for drainage design are covered in Chapter 11.

13.4 Fire protection services

As outlined in Chapter 7, specific requirements for fire protection services should be developed in conjunction with the overall fire strategy. The design impact of incorporating such systems can be significant (e.g., sprinkler storage tanks, the need for secondary power supplies etc.), so early planning and engagement with the fire engineer and building control officer as well as the client and their insurers is essential.

Where required, fire protection is most commonly provided by traditional wet sprinkler systems (in line with BS EN 12845[160]), and sprinkler storage and allocating appropriate plant space is one of the major considerations. Storage can typically require tanks up to 185m^3 for standard hazard systems, but this may rise to higher volumes if the hazard classifications are higher. Alternative fire protection measures can be considered, such as dry sprinkler systems, gaseous suppression or water mist systems, but each should be evaluated for their specific performance requirements (e.g., consideration for how effective water mist is if used in conjunction with smoke ventilation systems), and whether these are being employed primarily for life safety or for property protection. It is always advisable to agree any alternatives as part of the project's fire strategy and with the consent of building control.

The presence of electric vehicles should also be considered in relation to fire protection measures, as it may be appropriate to consider incorporating enhancements to assist in mitigating risks (e.g., consideration of sprinkler head densities, water storage volumes, enhanced drainage provisions or sprinkler bulb temperatures).

Fire protection systems require specialist designer input to develop detailed system requirements due to their requirements for certification. Therefore, engagement at the earliest stage is advised in order to develop these details in conjunction with the fire strategy.

13.5 Vertical transportation

After evaluation of the requirements for lifts for a specific development — with reference to the internal planning and management of pedestrians, safety legislation, and firefighting and evacuation requirements (Chapters 4 and 14) — the lift design process should consider the types of lift available, in conjunction with the performance that these can achieve.

Lifts can vary in terms of their operation and drive, and can be suited to different applications, uses, and demands. Cost may also be a factor in the selection of a particular lift type, but the main options are:

- Hydraulic lifts
- Machine roomless (MRL) lifts
- Conventional machine roomed lifts

Other lift types exist, such as platform and cabin lifts, but are less common in car parks.

Having considered the lift type, key design parameters including number of floors served, floor-to-floor heights, and occupancy/usage will be key to determining the appropriate lift solutions. Selecting the quantity, size and type of lift will be hugely impacted by these design parameters and it is always best to engage with a lift specialist who can provide evaluation of lift traffic analysis to inform the correct selections. Advice is also available in the literature[30, 161–163].

13.6 Sustainability and energy position

It is necessary to ensure that sustainability and energy use remain at the forefront of the designer's thoughts as the detailed design stage progresses. This is not only to promote good practice design, but to ensure that any commitments made in the planning stage's energy strategy are delivered.

In the UK, an 'As-designed' Part L strategy update may be required (if the building falls into the scope of the Building Regulations Part L[33]), and it is therefore important that the various inputs that influence this have been monitored through the design stage to avoid any unwanted surprises at the end of the design stage.

In particular, key mechanical and electrical (M&E) aspects that are likely to have the greatest impact are:

- Specific fan powers
- Lighting/luminaire efficiencies
- Control strategies
- Integration of renewables (e.g., photovoltaics)

14 Fire considerations (detailed design)

14.1 Introduction to fire safety engineering

Chapter 7 provides general guidance on the fire considerations to be incorporated into the initial stages of a design. This chapter gives more detailed and specific guidance.

Prescriptive design guides such as BS 9999[31] or Approved Document B[66] give good advice but should not be followed blindly as they may only be relevant to common building situations. Where it has been identified that fire engineering principles should be followed, a suitably qualified and experienced fire engineer should develop the detailed fire strategy. The guidance in this chapter should be sufficient for simple and common building situations, but it will always be for the designer to assess whether additional measures are required for their particular situation.

14.2 Risks from fire

It is important that the risks from fire are considered early in the design. The recommendations in this chapter are relevant to regular multi-storey car parks, underground car parks or those within other buildings. Risks can be to life safety, property or the environment and all should be considered.

Some specific risks from car parks arise from the materials used to manufacture modern cars, which often involves significant quantities of plastics. These can produce large quantities of toxic smoke, which most design codes do not explicitly consider.

There are different ignition risks from different vehicles. For vehicles with internal combustion engines, arson is often the greatest risk. Electric vehicles do not have quantities of liquid fuel to ignite, but can suffer from thermal runaway. This is worse during charging or for vehicles which have suffered damage. Vehicles which may be prone to damage from their use, such as police pursuit vehicles, may need to be treated with additional care, and this should be assessed by a fire engineer.

Consideration needs to be given to the risks of fire spread. Plastic fuel tanks may rupture, potentially resulting in a running fuel fire if the floor is sufficiently sloped. LPG and hydrogen powered vehicles, and electric vehicles may spread fire in a different manner.

When assessing risks, particular consideration should be given to car parks with 'sleeping risk' uses above them. Similarly, thought needs to be given to the risks from storerooms (either rubbish stores or general purpose) in car parks underneath residential blocks.

14.3 Means of escape

Users of the car park should be able to travel from any part of the floor plate to either a protected stairway, or to a final exit away from the building within the distances in Table 14.1. The layout of the points of exit need to therefore be within these distances.

Exits need to be sufficiently separated to prevent one fire blocking more than one exit. The first part of a user's exit route may be in a single direction for up to 25m — at which point the alternative exits must be at least 45° apart.

Table 14.1: Maximum distance to closest point of safety

Travel in one direction only	Travel in more than one direction
25m	45m

A single exit can only accommodate up to 60 people. Travel distances in open air may be extended with the advice of a fire engineer, as the risk is lower provided there is room to evacuate clear of any fire.

Exit widths should be no less than those shown in Table 14.2. This is measured as a clear width, free from obstacles.

Table 14.2: Minimum width of exits

Maximum number of people	Minimum width
60	750mm
110	850mm
220	1,050mm
>220	5mm/person

The number of people should be based on two people per car parking space, and one exit should be discounted as it may be unavailable due to the fire.

A place of relative safety has to be fire-separated from the potential fire compartment by at least a 30 minute standard of fire resistance (as tested to BS EN 13501[138]) and include doors fitted with smoke seals. From there, evacuation can continue either down or up the stairs, which should lead to a final exit away from the building. Final exits should not lead into areas where users cannot get away, such as courtyards.

Stairs should be sized using the following formulae:

$$W = P + 15N - 15 / 150 + 50N$$

Where:

$$P = 200W + 50\ (W - 0.3)(N - 1)$$

Where:

W = width of the stair (mm)

P = number of people using that stair

N = number of stories served by the stair

This is valid for all stair widths of 1,100mm and above, which should be considered the minimum width unless the number of people exiting is less than 50, the stair is not a firefighting stair, and the stair is not used for accessibility reasons. One stair should be discounted unless it is approached by protected lobbies, or a smoke control system conforming to BS EN 12101-6[164].

Fire engineering principles can be used to allow extended travel distances or reduced widths where there are mitigating features such as suppression systems and high ceilings. An Available Safe Egress Time/Required Safe Egress Time (ASET/RSET) comparison should be carried out by a competent fire engineer in this situation.

Consideration needs to be given to people with impaired mobility. This should be from all areas of a car park, not just accessible parking bays. Ideally, those with impaired mobility should be able to self-evacuate. Where lifts are provided, they should be considered as evacuation lifts with appropriate back-up power supplies and suitable refuge points. Where this is not possible, evacuation chairs should be provided within escape cores within staffed facilities, alongside a protected refuge with a two-way communication device so that a user can call for assistance. This creates an ongoing management requirement as staff will need to be regularly trained in the use of evacuation chairs.

Where a management regime is required, consideration should be given to the availability of staff for assistance, and the means of communication (via suitable refuge communication systems). Managed regimes should only be put in place where self-evacuation cannot be achieved.

Protected refuge areas should always be located in a smoke-protected area (i.e., after a fire-protected lobby). While it may be beneficial to provide non-protected help points within the main parking deck, these are not suitable for use in an evacuation. The size of refuge areas should consider the number of accessible bays on the parking deck being served. It can be beneficial to stack accessible bays over multiple levels, to prevent excessive demand being placed on a single refuge area, while also minimising horizontal travel distances. The practicalities of this should be carefully considered based on the size of the car park and the anticipated number of occupants, particularly for car parks that experience peak flows (e.g., serving a train station, sporting facility or music venue), as once an occupant has evacuated the building without their own mobility equipment they will continue to occupy the evacuation chair and will not be able to disperse without assistance. While assisted evacuation using an evacuation chair from a basement car park (i.e., upwards evacuation) is permitted by the building regulations, the health and safety implications for evacuation chair operators should be considered in the design. Powered evacuation chairs designed for assisted evacuation could be considered to mitigate this.

All exit routes must have appropriate fire exit signs, in accordance with relevant standards.

Automatic fire alarms should be provided which will assist in the early movement of people to a place of safety. In a building which has a phased evacuation strategy, a voice alarm must be provided, and the evacuation phasing must be in accordance with the fire strategy.

14.4 Structural fire resistance

Structural fire resistance is the ability of structural elements to carry the design loads, and should not be confused with combustibility or fire propagation. Some structural elements may also form fire compartmentation and will require integrity and insulation ratings.

Codified fire ratings can be used for common building types but should only be used where deemed appropriate. Sprinklers are an important feature in controlling fires and should be incorporated wherever possible. The need for life safety sprinklers can come from means of escape, fire resistance or compartmentation requirements, and sprinkler systems should not be omitted from car parks where the building has that requirement. Sprinklers are important for property protection and may be an insurance requirement.

Minimum structural fire times, when tested in accordance with BS EN 1363-1[165], are given in Table 14.3.

Where a car park forms part of another building, the fire time should be appropriate to the fire time of the building as a whole.

UK guidance allows for 15 minutes' fire resistance in car parks with a floor up to 30m high, where it is constructed by materials of A1 combustibility, and has permanent ventilation of 1/20 of the floor area spread around the facade. This should be adopted with great caution, as the car park may be susceptible to locally raised temperatures in fire if the layout prohibits good venting, even if it meets the guidance. Where this is planned to be used, it is recommended that a fire engineered solution is adopted using IStructE guidance[166] including a Qualitative Design Review. The guidance provides information on various measures that can be adopted – including time equivalence, which explicitly takes account of ventilation, along with other measures such as sprinklers.

It is important to consider the reaction to fire of the structure itself. Materials such as concrete have levels of fire resistance built in with appropriate thicknesses of elements, and the right levels of cover to reinforcement. Other materials such as steel require applied fire resistance. It is important that any applied fire resistance is robust enough for the location, is durable for the life of the structure, and that appropriate inspection and maintenance is carried out. Timber can be used as a structural material but care must be taken because the timber will contribute to the fire itself. Further guidance can be found in *Introduction to structural fire engineering*[79].

Table 14.3: Minimum structural fire times

Depth of basement floor		Height of building's top floor			
≤10m	>10m	≤5m	≤18m	≤30m	>30m
60 minutes	90 minutes	30 minutes	60 minutes	90 minutes	120 minutes

14.5 Compartmentation

It is recommended that the car park area is separated by fire-resisting construction from other uses such as retail areas, offices etc; and from rooms with ancillary uses such as plant. It is also recommended that cycle storage is compartmented separately from car parking.

The standard of fire resistance should be 60 minutes, unless the car park is larger than 450m^2, in which case it should be no less than the structural fire time for the building (with a minimum of 60 minutes).

For car parks with a floor >30m above ground, each floor should be a compartment floor with a 120-minute fire resistance. A means to stop fire spreading from floor to floor up the access ramps is required. Similarly, basement levels should be fire separated from the above-ground levels, and each floor in a multiple basement car park should be a fire compartment floor.

Enclosures for high voltage equipment, the perimeter of a firefighting core and enclosures to other life safety equipment must be to a standard of 120 minutes' fire resistance.

Fire doors to these enclosures should be to a FD60S standard, except to any areas where oil fuel is stored — which should be FD120S.

Power distribution networks may impose higher standards on doors and enclosures to network substations.

Penetrations through fire-rated construction must be adequately fire-stopped.

Care must be taken of details such as expansion joints, drains and plastic goods to ensure they do not compromise the fire resistance. The direction of fall should generally be away from the lift and stair cores to prevent burning fuels from reaching these zones.

14.6 Firefighting

Firefighters require access to a site, around the building and (where the building is tall or large) safe access within the building.

Access roads to the building should be suitable for fire service vehicles, including where appropriate high reach appliances. This information should be gained from the local fire authority. Water supplies for firefighting need to be considered, which may include public fire hydrants within reasonable distance, ponds or rivers depending on local needs.

Firefighting shafts should be provided (Table 14.4). Firefighting shafts generally contain protected stairs, ventilated protected lobbies, dry or wet rising or falling mains and in certain circumstances a firefighting lift. The specific requirements will depend on local firefighting processes. In the UK, BS 9999[31] should be followed.

Table 14.4: Firefighting shafts

Building type	Facilities to be provided
Top floor 11–18m above ground with all floors <900m^2	At least one protected stair with a dry rising fire main.
Top floor 7.5–18m above ground with a floor >900m^2	At least two firefighting shafts without the need for a firefighting lift.
Top floor above 18m with a floor <900m^2	At least one firefighting shaft including a firefighting lift.
Top floor above 18m with a floor >900m^2	At least two firefighting shafts including firefighting lifts.
Basement with lowest storey >10m below ground	At least one firefighting shaft without the need for a firefighting lift.
Basement with two or more stories >900m^2	At least two firefighting shafts including firefighting lifts.

Firefighting shafts should be located such that all parts of the floor can be reached within 60m of the fire main outlet. Access to the firefighting shafts at fire service access level should be within 18m of the point at which the fire service vehicle can park, so as to ensure easy charging of the fire main.

Goods and service lifts should not be sited in firefighting shafts.

For buildings with a floor <11m above ground, no internal firefighting provision is necessary. However, that means good access around the perimeter of the building is necessary to conduct firefighting operations from the outside. For buildings with total floor area >24,000m^2 100% access to the perimeter is required. This can be decreased linearly to total floor area <2000m^2 where none is required. If this cannot be achieved, internal firefighting shafts will be required.

Firefighting water must be available to suit local firefighting arrangements. This could be in the form of public hydrants within a suitable distance, rivers or ponds and lakes. If sufficient volumes and flows are not available, suitable provision will need to be made within the car park site. In most urban areas public fire hydrants are provided at frequent locations, and these can be checked with the local authority or fire authority.

There should be a clear indication of floor level — which is visible in low light and smoky conditions — situated in every stair landing.

Basement levels should be provided with smoke venting to assist with firefighting operations. Natural smoke vents should be provided so as to effectively allow cross venting with a combined area of 1/40th of each basement floor area, evenly spread around the perimeter. Mechanical smoke vents must be automatically operating, provide at least 10 air changes/hour (determined by an assessment of the fire load) and operate as a whole system at 300°C for an hour for a sprinklered risk. Unsprinklered risks require higher temperature rating of fans, ductwork and dampers, assessed by the likely compartment temperatures. Smoke extract systems should not breach the necessary compartmentation lines.

14.7 External fire spread

The external walls and cladding of a building need to resist undue spread up the facade and from building to building. There are also requirements for fire rated elements close to firefighting shafts[31].

The term 'relevant boundary' is used to determine the point at which external fire spread is to be calculated. It is usually taken as the centre of a road or the site boundary but should not be taken as at the face of an adjacent building. Both buildings either side of a relevant boundary need to respect the same boundary.

The external surfaces of the envelope should meet the ratings in Table 14.5.

Where the building has a floor >18m and contains sleeping accommodation, the materials of the envelope should be rated at least A2-s1,d0 except for materials such as gaskets and sealants that are absolutely required and cannot be sourced to meet the rating. In England other materials were specifically banned following the 2017 Grenfell Tower fire. While the recommendations in this guide are good practice, note should be taken of any local regulations and specific bans relevant to the location of the car park.

In any other building with a floor >18m above ground, the materials should meet A2-s3,d2 with the exception of gaskets, sealants etc.

Table 14.5: Fire ratings for external surfaces

Building type	>1m from relevant boundary	<1m from relevant boundary
Any building with either seven or more stories or a storey >18m above ground, which contains sleeping accommodation	A2-s1,d0	A2-s1,d0
Any other building with a height >18m above ground	B-s3,d2	Up to 18m: B-s3,d2 Above 18m: C-s3,d2
Any other building with a height <18m above ground	B-s3,d2	No requirement

Green walls have an important role to play in absorbing CO_2, dust and pollutants, and can enhance local habitats and biodiversity. They should be encouraged, but it is important to note that they can be a fire spread risk, particularly when the moisture content is <45%. Green walls can never be rated as class A, and so are not acceptable on buildings with a floor >18m containing residential accommodation. However, with appropriate selection of plants and substrate, and adequate irrigation and maintenance they can be used on other types of taller building[167].

Buildings which are close to their boundary run the risk of spreading fire to adjacent buildings and so may need to have elements of fire-rated construction. Calculation of the amount of protected area required should follow the methods in BR 187[168].

There is also a need to prevent fire spread over roof coverings, which would include the top deck of an open car park. Roof coverings should meet the ratings in Table 14.6, based on the distance from the roof to the relevant boundary.

Table 14.6: Fire ratings for roof coverings

Distance to relevant boundary	Rating
<6m	B roof (t4)
6–20m	C roof (t4)
>20m	F roof (t4)

14.8 Management

Fire safety information should be given to the 'responsible person' when the project is complete. The responsible person is defined in UK law, but in general is the person who is responsible for the operation of the building and its fire safety precautions.

Basic information on the location of fire protection measures may be sufficient. As-built plans of the building should be provided showing:

- Escape routes
- Location of fire-separating elements including fire doors
- Locations of fire safety services; including fire alarms, emergency lighting, fire safety signage, sprinklers and firefighting equipment

Any high-risk areas should be clearly highlighted, and details should be provided of all of the following:

- Specifications of any fire safety equipment provided, including mode of operation and routine maintenance schedules
- Any assumptions regarding the management of the building in the design of the fire safety arrangements
- Any provision enabling the evacuation of disabled people, which can be used when designing suitable personal emergency evacuation plans

The fire safety strategy should be provided, so that it can be periodically updated as alterations are made to the building.

More complex buildings, or those where fire engineering principles have been used should have more detailed information.

Further guidance is available in Clause 9 and Annex H of BS 9999.

This forms part of a golden thread of information which is needed throughout the building's life, to enable it to be appropriately managed and maintained. The information should be treated as a live document to be updated when alterations are made to the building or to the fire safety arrangements.

The fire strategy will include assumptions on the level of management required, and it is essential that the responsible person is competent — ensuring all other staff who have a role to play are adequately trained and supervised.

15 Inclusive design (detailed design)

15.1 Design of accessible parking bays

The pedestrian horizontal travel distance between accessible parking bays and the facility the car park is serving should be minimised — generally not exceeding 50m.

Within larger car parks, stacking accessible bays across multiple levels served by a lift is recommended to minimise horizontal travel distances for a large number of bays. This has the added benefit of reducing the size of protected refuge areas required on a single level.

For car parks with more than one entrance, e.g., serving different parts of a large retail complex, accessible bays should generally be evenly split between the different entrances.

Disabled users may need more time to access their vehicle, and may use space outside of the hatched access zone while doing so. For this reason, it is recommended to locate accessible parking bays in less busy areas and in cul-de-sacs to avoid obstructing vehicle or pedestrian access routes, as long as suitable travel distances to lifts, ticket machines and other important facilities are minimised.

Accessible parking bays should be located on a level surface to allow comfortable access to the vehicle. A gradient of 1:60 can be treated as level for accessibility purposes, and will allow for drainage/run-off requirements.

Standard accessible parking bays should be 2,600mm x 5,000mm, surrounded by a 1,200mm access zone on both sides and to the rear of the vehicle. Access zones may be shared between adjacent bays, provided each vehicle has 1,200mm access on both driver and passenger sides.

It is essential for the entire access zone to be clear of structure or any other obstruction, due to the variety of ways that wheelchair accessible vehicles (WAVs) may be accessed, and to be clearly demarcated from the carriageway. It is beneficial for the structural grid to be designed with this in mind, minimising internal structure wherever possible to allow flexible placement and conversion of accessible bays, as well as improve visibility and safety for all users.

A large accessible bay, 4,800mm x 8,000mm and with a headroom of 2,600mm, can accommodate late converted commercial vehicles or accessible minibuses serving groups of disabled people. This may be appropriate within car parks serving facilities likely to attract this kind of user group, e.g., leisure facilities or some public buildings. Due to the practicality and structural implications of providing these within a multi-storey car park, they may be provided on the ground floor or immediately outside the parking structure.

Designated accessible parking bays may benefit from additional lighting, as users may spend longer at the bay, e.g., transferring from or loading/unloading mobility equipment.

Generally, entering and exiting the vehicle onto a surface level with the parking bay is preferred, and this should be the case for all designated accessible bays. However, some side-entry WAVs may be easier to access from a standard 125mm kerb due to the gradient of a ramp, and standard vehicles can be easier for ambulant disabled people to access from a kerb due to a reduced stepping distance. For this reason, some kerb-access bays may improve accessibility overall, but should not be counted within the accessible bay provision.

15.2 Speed calming measures

The use of speed humps or rumble paving to calm traffic within car parks can cause particular discomfort to disabled users. In addition, WAV designs commonly require the floor of the vehicle to be much lower than a standard vehicle, such that speed humps can cause damage to the underside.

As such, these measures should not be used on vehicular routes to accessible parking bays. Barriers and route narrowing using bollards are generally preferred to calm traffic.

If speed humps are used, these should not extend into pedestrian routes as they can act as a barrier to wheelchair users, be difficult for ambulant disabled people to navigate, and present a trip hazard to all users.

15.3 Accessible EV charging

Where EV chargers are provided, an equitable proportion of charging points should be provided for accessible bays.

It is recommended that these are provided in addition to the standard accessible bays and EV charging bays required, to prevent people requiring EV charging facilities from using accessible bays inappropriately (or vice versa) due to a lack of general provision.

In general, accessible EV charging bays should be designed to the same requirements and principles as standard accessible bays with regard to travel distance, adequate lighting etc. However, a wider access zone between vehicles (1,600mm), and an additional clear access zone at the front of the vehicle, may be required to enable wheelchair access to the charging station, depending on layout.

EV charging stations associated with accessible parking bays should be accessible to a variety of users. The electric vehicle accessible charging document PAS 1899[169] contains principles that support provision of this:

- Clear visual indication of whether a charging point is available or out of order, that can be seen from a distance and the type of charging station (e.g., connector type/payment options available) so people can make decisions early in their journey about whether to use the parking bay (i.e., not required to park to read instructions)
- Preferably located centrally in front of the vehicle where there is sufficient space for wheelchair users to access the front of the vehicle and without allowing the vehicle to protrude into the driveway. Where this is not possible, the charge point should be between the vehicles at the end of the hatched area with a clear indication of which bay it relates to. If located side on, this may require a wheelchair user to leave a large space between the charging station and the vehicle to transfer to a wheelchair from the driver/passenger seat, compromising the available parking space or intruding on to the adjacent bay or carriageway, so this is not recommended unless there is a very wide bay
- An enlarged access zone of 1,600mm between the vehicle and EV charging station, to allow a wheelchair user to turn between vehicles. An access zone of 2m will allow access for larger powered wheelchairs and mobility scooters
- The charging station should be installed on the same level as the parking bay, so users are not required to navigate a kerb or upstand or take a longer/more complicated route to access the charging station
- Bollards or wheel stops should not impede access to the charging station. Locating charging station elements (e.g., docking station; screen etc.) on the front face of the charging station (e.g., facing vehicle) may be easier to access and view than if on the side
- Charging cables should not encroach into or block adjacent parking bays and/or access routes as it could present a trip hazard
- When the vehicle is charging, the clear width of the access route should remain sufficient for wheelchair users (e.g., trailing cables should not obstruct the clear width)
- The height of EV charge/docking points, controls and screens should be visible and reachable for seated users. Preferably, users can choose to reach from either a head-on or side-on position
- Controls should be operable with one hand
- Cable should be as lightweight and flexible as possible, and easy to grip and operate with one hand
- Thermally conductive surfaces should be avoided on all operable elements
- EV charging stations that require users to have a mobile phone or download an app to use can present extra barriers to use and should be avoided
- Within staffed car parks, a means of requesting assistance (e.g., via a call button or two-way communication device) is preferred

15.4 Wayfinding and signage

Clear, obvious and accessible directions, information and instructions for using the car park and associated facilities should be provided. This includes:

- Signage that is accessible to users at both standing or seated height. Lower-height signage should not pose a potential hazard to users (e.g., collision)
- Clear information regarding parking charges and height restrictions, evident at the entrance. Directional information should also make clear to motorists how to immediately exit the car park if they do not wish to park
- The availability and location of accessible bays (including those with accessible EV charging stations) should be signposted from the entrance
- Accessible pedestrian routes, particularly where these differ from non-accessible routes, should be signposted on all levels and from the immediate vicinity of accessible bays
- Considering the simplicity and consistency in signage (e.g., consistent use of symbols, typefaces, colours etc.)

Digital wayfinding information that is accessible to users prior to using the car park can help reduce the anxiety experienced when further away from home and/or in unfamiliar places. For example, this may include:

- Online maps and routes
- An overview of accessible facilities
- Information on the number of accessible bays and EV charging stations
- Real-time information on availability of accessible parking bay spaces and accessible EV charging stations

Signage should provide easy to follow steps on how to use equipment (e.g., ticket machine and/or EV charging station). Illustrations may help people who are either dyslexic or have lower levels of literacy for other reasons.

15.5 Pedestrian comfort

The comfort of a diverse range of pedestrian users should be considered in the design of lighting, acoustics, wind mitigation, and heating systems.

For example, reverberant acoustic environments can cause particular distress to some neurodiverse people, limit communication for people with different levels of hearing, and make it harder for people with reduced vision to navigate an environment[170].

The design of lighting and visual contrast should clearly identify safe routes and hazards, while avoiding areas of shadow or excessive contrast that may be perceived as steps or be distracting to both pedestrians and motorists.

In open-sided multi-storey car parks, wind mitigation may be required around accessible bays and routes, as high winds or unpredictable gusts can pose a particular hazard to people using mobility equipment, ambulant disabled people, or children.

Transitional spaces, e.g., lift lobbies, should avoid sudden changes in lighting, noise, or temperature.

Designing for natural surveillance, with clear sightlines and consistent and good quality lighting, can improve actual and perceived safety for a diverse range of users.

15.6 Ticket machines and barriers

Where ticketing/payment machines (devices) or barriers are installed, they should be accessible and usable by everyone. Motorists may expect to be able to access and use the machines without leaving their vehicles, so they need to be accessible from both inside and outside the vehicle.

Machines should be clearly identifiable, and sign-posted, throughout the car park. Where practical, they should be located close to accessible parking bays, to minimize travel distances for these users.

Any screens, controls, dispense points etc. should be able to be accessed comfortably by both standing and seated users, or vehicles at a range of heights.

Devices may be operated by people with varying manual dexterity, mobility and/or with limb loss. For example, a payment card slot and ticket remove slot may require users to make a pinching action to retrieve which can be difficult for some people; having to bend down to collect coins may be difficult for people with limited balance. Contactless methods of payment and ANPR cameras as an alternative to physical barriers are preferred where possible. The availability of staff assistance can mitigate many of these risks.

The ability for a diverse range of users to see and reach different elements should be assessed in relation to potential barriers. For example, the position of a bollard may restrict the viewing angle for a wheelchair user, causing the display to be obscured by glare.

Devices that require long sequences of actions or requiring the user to submit a vehicle registration number can present a barrier to people with cognitive requirements who may struggle to remember information. Devices that allow users to choose a photo of their vehicle taken via CCTV/ANPR on entry to the car park is ideal.

When exiting the car park, payment/ticket validation at an exit barrier should minimize the need for a motorist to reach outside the vehicle. While staff assistance can significantly mitigate this, an ANPR system that avoids a physical barrier is preferred.

Payment via an app or other online method can be more accessible to some users, by minimising the need to travel to a physical ticket machine or ticket validation point. Users of assistive technology may also find their own device much easier to use than a ticket machine or unfamiliar interface. However, this method should not be relied upon as the only means of payment, as it may disadvantage users without access to or experience of using technology, for example. Where an app-based system is used in an underground or multi-storey car park, mobile phone/Wi-Fi hotspots may be required.

15.7 Staffed facilities

In general, staffed facilities and the availability of assistance improve accessibility.

In the design of staffed car parks, the availability of assistance can be maximised by considering the visibility of key areas including lifts, barriers, payment machines and accessible bays, from the staffed area.

A means by which to contact staff for assistance (e.g., two-way communication system to a ticket booth or control room) can also be provided in lift lobbies, EV charging stations, or rest/help points. Communication systems should include both verbal and non-verbal and visual and non-visual means of communication (e.g., a video intercom).

15.8 Sanitary facilities

Accessible sanitary facilities may be beneficial where accessible EV charging stations and parking bays are provided, and users are required to wait as the vehicle is charging. Disabled people, older people and children may require the use of sanitary facilities more often.

If sanitary facilities are provided for car park users, this should also include accessible facilities. Further information on the design of accessible sanitary facilities can be found in BS 8300-2[25].

Where only one WC is provided, it should be wheelchair accessible.

15.9 Inclusive means of escape

As a general principle, independent evacuation for anyone using the car park should be the aim. Where this cannot be achieved, safe refuge points where a person can call for assistance with their evacuation may be provided.

Step-free evacuation should be considered essential from all levels of a multi-storey or underground car park – it cannot be assumed that people requiring step-free access will only use designated accessible bays. Further guidance is available in Chapter 14.

16 Suicide prevention

16.1 Introduction

Organisations that design, manage, or operate multi-storey car parks should consider the issue of suicide prevention, response, and recovery.

2021 statistics[171] show that in England there were 5,219 deaths by suicide (307 more than in 2020). In Wales there were 347 (62 more than in 2020). In Scotland there were 753 (52 fewer than in 2020) and in Northern Ireland there were 219 (14 more than in 2020). Each devolved nation has a suicide prevention policy with the key aim to reduce the suicide rate in the general population and better support those bereaved or affected by suicide.

Part of the suicide prevention strategy is to restrict access to the means of suicide. Multi-storey car parks are open structures, providing easy access to great heights, from which jumping offers a perceived higher certainty of death, according to suicide experts. Jumping is relatively easy, involves no cost, and lends itself to impulsivity. People considering suicide can be more easily deterred if office buildings and other high structures are closed at night or have security personnel or other measures in place, and have sealed windows and security in the lobby. Car parks on the other hand, often have open sides for ventilation, allow customers to enter with little or no security checkpoints, and generally have fewer people walking around, making them much more accessible.

Suicide prevention is important from a moral standpoint as well as to minimise trauma to parking professionals and other witnesses or bystanders.

A number of effective steps can be taken to prevent car parks from being used for suicide attempts and to increase the chances of last-minute intervention. Many of these are aligned with the Safer Parking Scheme and Park Mark® Award, such as CCTV or foot patrol surveillance, good lighting and a clean environment[17].

Suicide risk itself is not specifically dealt with in the Building Regulations or statute other than to say that as described previously, the building should be safe for use by all and that there is no clear exclusion of this rule to those who are intent on committing suicide.

Note, that the recommendations within the Approved Documents relating to guarding are based on the requirements to prevent accidental falling from the building and not necessarily to prevent intentional acts.

Designers can seek support when considering the risk of suicide (and indeed the topic of suicide prevention) from Designing Out Crime Officers (DOCOs) who are members of police staff specialising in the act of mitigating the ability for a crime to occur within a building or public space via the design of that particular built environment (Section 16.3).

This is an area where context associated with location of the car park and indeed the user group who will utilise the car park is of significance. For example, if a car park is to be located within a hospital area, especially if that hospital has a unit specialising in mental health, it is far more likely to be a significant risk in terms of acts of suicide or attempted suicide.

Potential measures to mitigate suicide attempts could include managing access, promoting and increasing visibility (both direct and indirect such as CCTV), providing emergency contact points and access to helplines, and improving the quality of the environment either aesthetically via planting, lighting or other decorative treatments or perhaps audibly by playing background music. More information on general security measures can be found in Chapter 9.

The utilisation of a roof over the top deck of a multi-storey car park can also improve the perception of safety and security, and the feeling of being enclosed can potentially dissuade those considering suicide from progressing. However, it is of paramount importance that all aspects of the structure be considered as a whole e.g., an edge protection impact barrier located adjacent to some large electrical conduit located adjacent to a stair core roof may result in a series of unintended access steps towards a platform at considerable height, which may lead to those intending to attempt suicide being enticed into utilising the building for that purpose.

It is generally accepted that those intent on taking their own life often plan the attempt in advance and are likely to use various forms of media to facilitate the attempt. It is therefore a prudent step to ensure that any suicide related stories or data published relating to a particular car park (or indeed car parks in general) are that suicide prevention techniques have been actively considered and adopted. This can help prevent the car park from being identified as a site suitable for suicide attempts in the first place.

16.2 Current guidance and responsibilities

Edge protection as discussed in Chapter 12 is for the prevention of accidental falling. The measures required to deter suicide, resulting from the risk assessment, are likely to be greater than that required to prevent accidents. The standard 1,100mm high guarding is relatively easy to climb.

In reality the more exposed the location of the required edge protection, the more inadequate the 1,100mm height seems to be. The difference in the amount of exposure one experiences between an upper deck of a MSCP and the top deck without a roof is considerable, yet the guarding height is the same. The designer should consider increasing the edge protection of vulnerable upper decks as well as top decks. This will reduce the opportunity to climb over the edge protection but will also break the visual connection with the possible fall to the ground.

16.3 Designing Out Crime Officers

Designing Out Crime Officers (DOCOs) are part of the community policing 'family' of police in the UK.

They provide specialist advice and guidance regarding the built environment at every stage of architectural design from pre-planning to the full development control process; minimising crime, disorder and anti-social behaviour, including suicide prevention.

DOCOs will liaise with architects, designers and developers — promoting an awareness of the Designing Out Crime service, National Police Chiefs' Council (NPCC) approved crime prevention initiatives, including Secured by Design[172] — to build awareness of the principles of crime prevention through design, and to encourage adoption of recommended standards and specifications.

As well as engaging directly with the designers, they will review planning applications to identify areas of risk and escalate these appropriately.

As DOCOs participate in problem solving forums with neighbouring forces to share best practice, benchmarking and issues relating to designing out crime, they can often add specific knowledge and experience to a design process which may otherwise not have been available.

Engaging with a DOCO with appropriate background and knowledge of suicide risk can be an extremely positive step towards mitigating risk of suicide through design.

16.4 Design considerations

As with any risk associated with a car park, the level of risk associated with suicide is dependent on a number of contributing factors. Some of these can't be reasonably foreseen at the time of design, such as future developments or changes in local activities, but all known contributing factors should be taken into account during the design process in order to inform the level of risk and therefore the level of mitigation required.

The likely intended user of the car park should be reasonably foreseen and taken into consideration. This will be a significant contributor to the level of risk associated, and is due to some user types being more susceptible to suicide risk e.g., those suffering from mental health issues and/or addiction.

Designers should not just look at the primary intended user group (i.e., people parking vehicles), as a secondary user group type may also use the building.

Consideration should also be given to the fact that the car park may be accessible to those not intent on using the car park for its primary function, i.e., the car park may be used or accessed by those not in either of the primary or secondary groups.

If a specific car park or even an elevation in a car park affords access to a surrounding area which itself carries a risk, for example an active train line or access to an adjacent building which in turn could provide an opportunity for a suicide attempt, additional edge protection should be considered in order to reduce the risk. It is not always the height of the car park deck and the user group which are key areas of consideration; the surrounding context also needs to be considered when the designer is identifying the risk and likelihood of potential suicide attempts.

16.5 Practical solutions

There are a number of potential methods available to designers to mitigate risk of suicide, ranging from physical barriers which may seem quite obvious, to more obscure and subtle measures such as increasing cross-deck visibility through circulation module selection.

The following is not an exhaustive list or a recommended best practice for every situation. Nevertheless, it is intended to provide possible ways a parking facility may address potential suicide attempts. The success of these measures depend on how determined the individual is, and their physical abilities. It is not recommended that car parks are turned into prison-like structures with 3m high walls topped with razor wire, but that a considered and appropriate response is made.

It is reasonable to design on the basis that the harder it is to gain access to the edge of the car park and a location to attempt suicide, the less likely it is that an attempt will be made.

16.5.1 Physical barriers

16.5.1.1 Fencing and screening

Installing physical barriers, such as chain-link fencing, screening, security netting (which could be less dense and more attractive), vinyl-coated mesh, metal grating, stainless steel or glass barricades, is recognised as a possible method to help prevent suicide. Where enhanced balustrading is required as a mitigation measure, full height infill mesh should be installed on the intermediate decks where there is a risk of serious injury from jumping from the car park, and on the top deck the height of the balustrading should be increased to 2,500mm above deck level with the top cranked inwards (Figure 16.1). To make it difficult to climb, it is important to select a mesh size that minimises the possibility of achieving a foothold.

Figure 16.1: **2.5m pedestrian guarding to assist in the prevention of suicide attempts**

An alternative is to construct a roof over the top deck and use full height infill mesh at roof level (Figure 16.2) and any levels below that are considered to be high risk.

Consideration should be given to the ease of climbing the barriers, and all reasonable attempts to ensure that the edge protection is not climbable should be made. For example, ensuring the apertures are kept small (certainly no greater than 50mm in any direction) and directing the top section of a fence (out of reach) inwards.

If there are parapets, barriers, or other structures near parking spaces that can be used to stand on, the height of the barrier should be increased appropriately.

Other elements that may pose a risk include steep ramps between floors where there are voids between the decks, open stairwells, and ventilation openings. Open areas can be sealed with treillage, grillage, metal screening, or stainless-steel cabling that permits air and light to circulate. If vertically mounted cable is used, it should be slanted inwardly toward the top of the parking ramp to make climbing more difficult.

One method of mitigating the risk of the car park being accessed by those not parking, would be to include access control equipment.

Figure 16.2: Full height guarding to underside of roof

16.5.1.2 Perimeter treatments

Soft landscaping (using trees, bushes, plants and grass) around the car park perimeter can be used to block the open landing area and also to reduce the perceived height of the building.

People tend to jump onto areas of concrete or hard landscaping; they don't want to see awnings or landscaping that could break their fall. Consider tables and seating if an area permits.

Where soft landscaping or trees can't be used, for example at vehicular entrances/exits, then perhaps the use of soft canopies or projecting signage attached to the elevation of the car park can be used to break up the vertical space between the ground and higher levels.

16.5.2 Geofencing

Other deterrents include geofencing (using CCTV) in order to monitor and secure specific areas of the car park. This can be set to alert the operator (usually via a message sent to a desktop or mobile device) at specific times, or if the system identifies someone in an area for an unusual amount of time.

Ultimately geofencing can be used to raise an alarm to the operator so that physical intervention can then be implemented.

16.5.3 Signage

Placing signage that promotes crisis intervention services, and/or sources of counselling in high probability locations can encourage individuals with suicidal intent to call the helpline number. It is important to note that there are differing opinions among suicide-prevention experts about using signage and the available evidence does not support placing signage everywhere. This is because crisis signs could draw attention and highlight a particular location as a method of suicide. Where possible it is advisable to use signs in conjunction with other measures such as staff training.

Signs could be placed in any location where it is felt that vulnerable people may see them and use them to reach out for help. There should be a live communications help point in the vicinity of the signage.

16.5.4 Visibility

Increasing visibility across the decks (and around all elements of the car park perimeter in particular) increases the chance of dissuading suicide attempts due to a lack of privacy (perceived or actual) and also allows quick intervention from staff or other users.

Visibility can be optimised by means of preventing any cul-de-sac locations or blind areas and also by adopting a car park layout which promotes across-deck visibility such as flat decks or vertical circulation modules (VCMs) to a lesser degree. Use of split-level decks or decks with large internal ramps with walls around them will break the deck up and reduce visibility.

Of course, indirect visibility can also be used either passively e.g., through the use of mirrors, or actively via CCTV.

CCTV systems are clearly more effective if they are being monitored in real-time, but prominent unmonitored cameras and signage may also help to dissuade suicide-related activity.

Some CCTV systems will have motion-activated cameras with monitors. Surveillance footage that captures a person displaying suspicious behaviour may allow for swift intervention by security or other staff.

Other measures include increasing the opportunity and capacity for human intervention, and increasing the opportunities for individuals to seek help by designing circulation routes and directions appropriately (refer to Table 2 of *Preventing suicides in public places: A practice resource*[173] for more detail).

16.6 Costs

All safety options have advantages and disadvantages. It is less expensive to plan for physical barriers when building a new structure than to retrofit an existing car park, so it's financially prudent to incorporate suicide prevention measures into the initial design.

Unfortunately, no matter what the extent of preventative measures, there is no guarantee of avoiding future suicides, because typically an entire car park cannot be fenced to a completely secure level. The goal is to provide a deterrent and reduce the risk as far as is reasonably possible.

There is of course a moral cost to doing nothing, in addition to the financial cost of having to wholly or partially close a car park for an indeterminate period following a suicide/attempt.

16.7 Legal issues/liability

Owners and operators of car parks have a legal duty to ensure that the building is safe for use. Liability should be a concern when a suicide/attempt occurs and while it is reported that actual incidences of parking facilities being sued are uncommon, the possibility of litigation remains.

17 Quality control during construction

17.1 General

The durability of a structure is often more sensitive to site practice than to the choice of materials used.

While general statements in specification clauses usually include these matters, the lack of quality control checks and adequate supervision on site can compromise durability of both prefabricated steel and precast concrete elements, with *in situ* concrete construction being more open to abuse.

Specification of concrete for durability is given in Section 11.10. Good practice for concreting works is covered in the literature[116, 131, 174–176]. In the severe exposure conditions to which car parks can be exposed, a higher quality of construction is needed to achieve a durable structure; this should include using a design that can easily and reliably be constructed to achieve good concrete quality and the minimum reinforcement cover. Similarly, the quality of handling and protection of steel members and their factory-applied protective coatings[107] will have a significant effect on long-term durability as onsite "touch-up" of chips and scrapes may not offer the same corrosion resistance as the factory-applied coating.

17.2 *In situ* concrete quality

17.2.1 Introduction

Car parks should not be considered as regular buildings, although they are commonly an integral part of a new transportation or commercial development. Commonly, car parks are in whole or in part external exposed structures, with the deck slabs and ramps suffering the worst exposure conditions. In the UK, this exposure can lead to conditions that are in some ways worse than for civil engineering structures such as bridge decks (Section 11.10), and they may need additional protection as a result.

For *in situ* concrete framing and slabs, despite all the care taken in designing the structure and specifying the minimum cover and concrete mix composition[13, 106], the final actions of installing the reinforcement and formwork, and placing and curing the concrete are equally essential to achieving the intended durability. A lack of attention to detail when either designing temporary works, or placing, compacting and curing the concrete, or prematurely stripping formwork, or overloading by construction materials and plant before the concrete has gained sufficient strength, can singularly or in combination negate the design efforts, resulting in a structure with poor durability.

For car park construction and for the slab and ramp decks in particular, the whole workforce should understand the objectives of these recommendations and use high standards of construction practice and maintain the specified tolerances[131]; for example, ensuring the correct falls are regularly achieved and will prevent water ponding, thereby minimising the risk of slip hazards in winter and obviating the need for operators to use extra de-icing salt. Good ventilation and drainage cross fall of at least 1:60 should mitigate the risk of ponding from water ingress and counteract the effects of deflection (Section 17.2.2).

Given the extra demands on building accuracy and concreting practice, adequate supervision should also be built into the contract to ensure the design intent is realised, particularly in situations where the exposure of the structure is severe.

17.2.2 Construction tolerances

The severity of exposure in car parks makes it essential to achieve the specified minimum reinforcement cover; constructing to nominal cover that allows for build tolerances. The reinforcement position must mirror the intended falls in the slab or ramp and thereby prevent depths of cover being either too low or too high[131]. The cover to reinforcement can only be consistently achieved if there is adequate tolerance on the reinforcement bending, the formwork and steel fixing, and if the reinforcement can be fixed as detailed.

As noted in Section 11.10.6, even where a waterproof membrane system is to be applied, the cover required for durability should not be reduced because of this protection. This is because the membrane has a finite service life and often the owner/operator of the car park will not replace the membrane until it has detached in substantial areas, or attend to leaking joints until there are complaints, by which time the concrete may be bare and saturated with salty water.

Before pouring concrete, checks are recommended to review the following:

- Drawings; to assess buildability before starting construction
- Formwork proposals
- Proposals for steel fixing and in particular the frequency of spacers
- That nominal cover is achieved, such that minimum and maximum cover limits are not exceeded
- Ensuring multiple overlaps do not reduce the cover, particularly where fabric reinforcement is used
- Proposals for placing and compacting the concrete
- Curing methods and provision for protection in inclement weather/overnight frost

For the top surface, the specified cover must be checked and the required datum level recorded by the contractor prior to concreting.

During placing, it is most important that the finished level of the top surface of the slab or ramp is placed accurately in accordance with the construction drawings. Failure to achieve the correct levels and falls can reduce the cover depth at the top, which is the most critical area with respect to durability (Section 11.10), or increase it unacceptably causing wider cracks. Details that may reduce cover thickness must also be carefully considered to prevent minimum cover requirements being breached, such as recesses for drainage, or cover to shear studs and other ferrous metal.

Deflection during and after the concrete pour is a particularly important consideration with composite construction e.g., precast planks with a concrete screed or steel-concrete composite decking, which may develop excessive mid-span deflections. Also, with a concrete frame, the precast prestressed units should not be placed directly onto the concrete bearing, as this produces high local stresses and can result in spalling of the cover on the supporting framing beams (Section 17.3).

Immediately after concreting, the top surface datum level must be re-checked and adjusted as necessary before the concrete stiffens. For formed surfaces, the cover must be checked immediately before placing the concrete, and spot-checks made after the formwork has been removed.

17.2.3 Ordering concrete

Section 11.10 provides guidance on achieving durability for the prevailing exposure conditions, which is the responsibility of the designer. The main contractor will commonly be responsible for ordering the concrete, which should be supplied by a quality-assured provider to the requirements of BS 8500-2[13] and Annex G of BS EN 206[105].

Contractor preferences on methods and rates of placing will affect the mix selection, and consistence will normally be determined at that stage, with trial mix designs provided by the concrete producer for approval by the designer and contractor. The contractor or producer may also advise on alternative mixes for frost protection, such as replacing a lower strength concrete with entrained air, with a higher strength mix. Changes at this late stage need careful consideration by the designer to ensure unintended consequences do not occur (e.g., higher creep and shrinkage).

At the point of delivery, the concrete producer will provide a delivery ticket for checking and sign-off by the contractor, and it is important that this process is supervised. The ticket provides an important verification that what was ordered was delivered, along with the time of mixing, transit time to site, and whether any water was added to the mix at any stage. If there are discrepancies between the ticket and the order, the mix should be rejected.

It is permitted for the driver of the concrete delivery truck to adjust the water content at the time of discharge, to ensure the specified consistence is achieved; evaporation during transport is the most common reason for extra water to be needed. The driver should record on the ticket the metered quantity of water added from the truck's tank, and the concrete should be thoroughly remixed. Concrete should be rejected if the consistency remains out of tolerance, such as being either too high or too low in consistence/slump.

The contractor should not order the truck driver to add more water to the mix to increase the consistence above that which has been specified and agreed. If extra water is added, it will proportionally reduce the strength and durability of the concrete. Also, while the mix will be easier to place on a flat slab, the concrete may not maintain the surface fall tolerances due to secondary flow.

As set out in BS 8500-2, concrete should be delivered not more than two hours after mixing. As longer transit times may affect the consistence and strength, loads that have been delayed by more than two hours should be rejected.

17.2.4 Placing and compaction

Best practice should be followed to ensure the concrete is placed carefully, using a tremie pipe to prevent dropping concrete from height or cascading over the reinforcement, as this encourages honeycombing and segregation. The mix must then be vibro-compacted to full consolidation and finished to achieve the correct falls, unless self-compacting concrete (SCC) is to be used.

When placing concrete onto ramps and decks laid to falls, greater long-term durability is achieved by starting the concrete placing at the lowest point and working up the slope, which reduces the risk of slumping, segregation and tearing of the concrete while in a plastic state.

SCC is being used increasingly for a variety of structures including car parks (Figure 17.1). It can offer significant benefits in terms of the speed, quality and cost of construction, particularly with congested reinforcement, but requires special consideration in terms of formwork design and the requirement not to vibrate[177]. While SCC is ideal for foundations and structural framing, special considerations are needed for slabs and ramps due to the self-levelling properties of the concrete. It will not be possible to achieve the designed deck and ramp falls without a further operation (e.g., a bonded non-structural concrete screed[115]), which creates other risks, such as delamination of the screed due to workmanship issues. SCC is included in BS EN 206 (Annex G) and BS 8500-1[106], with further advice available in the Concrete Society's review[177].

For compliance purposes, the air content of air-entrained concrete[178] is usually checked by the producer before dispatch, and then measured again at the point of delivery. When the concrete must be conveyed over long distances, particularly by pump, changes in the air content can occur between the point of delivery and the point of placing:

- If the air content at the point of compacting is too low, this will reduce the effectiveness against freeze–thaw action — mitigate by increasing the admixture dosage

Figure 17.1: Self-compacting concrete

- If the air content becomes too high, this is also important as the compressive strength and density of the concrete will reduce — mitigate by reducing the admixture dosage

The air content at the point of placing should be established at the beginning of each concrete placement, as well as each time the placing conditions change.

17.2.5 Bleeding and plastic cracking

Immediately after vibro-compacting, the concrete surface will develop a sheen, as bleed water slowly escapes from the mix. The process of bleeding is common with standard concrete mixes and helps prevent plastic shrinkage cracking. Excessive bleed of the concrete can cause plastic settlement cracking in the surface. Both plastic shrinkage and plastic settlement occur while the concrete is still in the plastic state (Table 17.1)[81]. With supervision, the incidence of cracks can be identified and action taken immediately to close them by re-vibrating the concrete, provided the timing is carefully considered.

Table 17.1: Types of cracking as a result of concrete bleeding

Plastic shrinkage cracking	Plastic shrinkage cracks form where the rate of bleed is less than the rate of evaporation from the surface, resulting in surface tension that splits the soft (plastic) concrete. Plastic shrinkage cracks can run deeply into the concrete and are not controlled by the reinforcement. Prevent by good curing practices, counteracting high air temperature, low relative humidity and high wind speed across the concrete surface.
Plastic settlement cracking	Plastic settlement cracks form when excessive bleeding of water from the concrete occurs due to sedimentation (the dense cement and aggregates sink, and excess water rises). This results in a slight settlement of the concrete surface. If there is restraint to settlement, such as reinforcing bars, the concrete is prevented from settling and cracks form. Prevent by using properly graded aggregates and low water demand.

It is not acceptable simply to finish the surface with a tool to drag laitance into the plastic cracks, as they will soon be exposed again by the wearing action of vehicle tyres and will form a direct path to the reinforcement. Guidance on non-structural cracks and their prevention is given in the Concrete Society's *Non-structural cracks in concrete*[81] and for information on repair refer to Section 11.10 of this guidance.

High strength concretes with a low water/cement ratio, and those containing entrained air or silica fume need extra care during placing. These mixes may have higher cohesion or thixotropy and to achieve full compaction, longer vibro-compaction times may be required. In addition, because these concretes have little bleed, there is less natural protection against plastic cracking and extra protection will be needed (Section 17.2.6).

Particular attention should be paid to achieve full compaction adjacent to construction and expansion joints, because any subsequent water penetration at joints will affect the durability of the structure.

17.2.6 Protection and finishing

A compromise is often required between the needs of curing, protecting and texturing the concrete surface. If done properly, protection and finishing of the concrete surface will be a labour-intensive operation. General guidance is available in the literature[116, 131], but they give no specific guidance for car parks.

The timing of the protection and finishing is of fundamental importance if a dense and durable slab top surface is to be produced. The finishing operation requires a concrete surface that has begun to stiffen, yet is generally free from bleed water (Section 17.2.4). If the surface is textured too early, bleed water will be mixed into the top of the concrete, making it more workable, reducing the texturing and reducing the wear resistance of the surface. Rainfall will have a damaging effect on the concrete surface, and the works must be adequately protected; rain not only forms pockmarks on the surface but as with bleed water, it will mix into the surface during finishing.

Finishing requires particular care to accommodate the sometimes conflicting needs of providing good skid-resistance, the application of a waterproofing membrane and the provision of an unimpeded surface for water run-off and dissolved salts (Figure 17.2).

Figure 17.2: Surface finish

During the finishing operations, surface evaporation from the slab should be minimised. This can best be achieved by avoiding extremely hot, dry or windy weather conditions for placing slabs. Where practical, the slab should be sheltered from direct exposure to the sun and rain, and protected against the passage of wind over the surface.

Once textured, the surface must be protected from severe drying conditions if plastic shrinkage cracking is to be avoided. Slab and ramp construction is particularly prone to plastic shrinkage cracking, where airflow can be concentrated over the surface.

The final finishing process for 'as-finished' concrete surfaces can be a subject of potential disagreement. Appearance of brushed and tamped finishes are particularly subjective and little guidance is in place for car parks, though TR 66 gives recommendations and photographs for hardstandings[119]. The surface should be defined as a 'special' category and accurately described in the contract documents in terms of the initial regularity of the surface and the tooling to be used to create the finished appearance (Section 11.10). A lightly brushed finish gives better durability than tamped finishes, as the latter is more operator-sensitive and can increase the risk of bleed water intermixing, producing a less durable surface.

Traditionally, brushed or tamped finishes were drawn across the concrete surface in the direction of the fall, allowing water to drain from the formed grooves in the surface. Experience has shown that it is better to run the grooves in the shorter direction i.e., between the main beams and movement control joints (typically 7.5m centres) as there is better control of the workmanship for finishing.

Provision should be made for the contractor to prepare a trial panel to set an agreed standard for the finish.

For concrete surfaces to receive liquid-applied membranes (Section 11.10.6), the manufacturers will usually require a smooth, closed texture, but not one that has been power trowelled to a polished surface. This is commonly described as a 'matt trowelled' surface, requiring only a light power trowelling (minimal passes, blades at low angle) to remove slight undulations and steps in the surface, before further treatment is carried out.

17.2.7 Movement provision

The structural design will identify main movement joints in the slabs, often with a requirement for intermediate (saw-cut) joints at typically 6-8m centres in bonded concrete screeds or ground-supported floor slabs, to prevent random drying shrinkage cracking.

Where saw-cuts are specified, the cut is usually made using a 4mm blade road saw approx. 18–24 hours after casting, by which time the concrete is strong enough to resist the saw blade damaging the surface by plucking out coarse aggregate pieces. The saw-cut depth is typically specified at ¼ to ⅓ of the concrete thickness. With reinforced concrete, care is required to position the reinforcement fabric or bars at an appropriate depth, to ensure the saw blade misses the steel, and that the correct minimum cover is maintained. Sawn joints should be sealed with a bead of mastic.

Proprietary joint systems are commonly specified for movement joints in car park roof decks, often requiring a recess let into the concrete into which the unit can be housed, and these joints are not discussed in detail here. Owing to the requirement to place the membrane under the joint, these units are usually one of the last items to be installed.

At intermediate deck levels, the movement joint may only require a mastic sealant to prevent run-off from cars leaking to lower decks, but as with all mastic joint systems, the sealant needs regular inspection and replacement as required, particularly in the first few years when drying shrinkage of the frame will be greatest.

17.2.8 Curing

The resistance of concrete to both chloride ingress and carbonation is substantially enhanced if the cover layer of concrete has fully hydrated in the period following casting. To achieve this, full curing measures must be taken immediately after the finishing is complete.

Concrete containing CEM 1 cement hydrates rapidly and good surface qualities can be achieved with 3–6 days of curing that prevents water loss from the surface. Commonly, concretes contain blended cements that contain fly ash or blastfurnace slag (broad designation CEM II or CEM III), with the common cements available in the UK listed in BS 8500-2[13]. Blended cements need extended curing for full surface hydration, of perhaps 6–10 days, for resistance to wear and carbonation to be achieved.

Curing should begin immediately following completion of finishing operations for each area[116]. Field experience has shown that traditional methods of curing, such as laying wet hessian on the slab covered with plastic sheets for a minimum of seven days, produces good results for concrete cast *in situ*, particularly where a rough, tamped or textured surface is used. The hessian must be kept damp during this period by adding potable water if required.

High performance spray-on curing membranes are available that have a similar performance to traditional wet curing and are particularly suitable for smooth power-finished surfaces (Section 11.10). If curing membranes are used, care is needed to ensure the product is applied at the correct application rate, and that losses carried away by wind are avoided. The curing agent must also be compatible with any waterproof membrane to be applied subsequently (Section 11.10.6).

17.3 Steel framing

Where the design uses a steel frame, the common options for finishing the ramp and deck areas are with precast concrete units and an *in situ* concrete wearing screed, and concrete composite decking (Sections 17.3.1 and 17.3.2). The design process will have identified the corrosion protection (Section 11.10) and fire protection requirements, allowing the framing to be fabricated, painted and protected ready for careful assembly. The process of steelwork installation is not considered here, other than to emphasise the importance of careful handling to protect any factory finishes. The main issues concern the choice of slab and ramp surface.

17.3.1 Precast concrete units

Solid or hollow-cored precast prestressed units are commonly specified for car park construction, allowing for rapid build with reduced workmanship requirements compared with *in situ* concrete. The units are typically pre-cambered and incorporate an *in situ* concrete wearing screed, such that when the screed mix is applied, the pre-camber is removed, and the plank surface is left level.

Careful consideration is required when specifying deck falls and ramp gradients, as the concrete screed mix should be laid at constant thickness, and not laid to a variable thickness to create the falls; changes in the thickness of an *in situ* concrete screed is likely to cause cracking, curling and de-bonding issues. Therefore, the steel framing needs to be set to the required falls, so that the planks can be installed, and the screed applied to give a surface with the required fall.

Planks aligned together will usually form a pocket between them that may be reinforced, and should be filled with an appropriate high flow, high strength grout mix. Commonly, pre-bagged non-shrink proprietary grouts are used for this purpose, mixed with a fine pea gravel (<6mm size). As the gap at the bottom of the plank can be variable or edges may be damaged, it may be necessary to use a liner or expanding foam to stop the grout escaping. At the butt joint, seated over the steelwork frame, dowel reinforcement will be provided to prevent differential curling. This location is particularly prone to movement, caused by shrinkage contraction of the screed and precast plank, and flexing of the plank under vehicle loading, which will cause cracks in the surface.

Guidance on the workmanship requirements for the installation of bonded screeds is given in BS 8204-2[115].

Waterproof membranes are commonly reinforced with a glass-fibre fabric over all joints between the planks to prevent the membrane from splitting (Section 11.10.6).

Temporary or demountable car parks can be designed and built with no concrete screed, and just a grout used to seal between the precast planks, followed by a waterproof membrane system. The performance will not be as assured as a slab with a bonded screed, as any slight movement of the planks under load, or from environmental changes (differential expansion and contraction), is likely to split the membrane unless it is fully reinforced.

17.3.2 Concrete composite decking

In building construction, steel framing with galvanised steel decking and an *in situ* concrete is a very common and rapid form of construction. However, for use in car parks, the design and construction need very careful consideration and may not be appropriate without special measures[126].

The concern arises from the penetration of water through plastic and drying shrinkage cracks that commonly form in the *in situ* concrete. Under flexure from cars, the cracks widen and extend and can reach full depth, down to the steel decking. Structural and durability design can define cover and concrete quality requirements to prevent reinforcement corrosion (Section 11.10), but corrosive attack of the structural steel decking is outside the code requirements. If it were to occur, it would progress largely unseen on the inside face, resulting in loss of section and reduction in structural capacity.

Mitigation measures against corrosion of the steel decking include sealing all cracks and interfaces, such as slab/column junctions, to block all possible paths for moisture percolation, and then providing a reinforced waterproof membrane over the deck surface. There remains a residual risk that drying shrinkage, and thermal and structural movement over the first few years may cause new cracks to form, and so regular inspection and maintenance is required during this time. Metal decking systems are available that have extra thickness of galvanising, or are pre-primed with a protective coating to resist corrosion, but their long-term performance is not known. As a result, even standard rainwater run-off has the potential to penetrate through the concrete and collect at the interface between the concrete and steel decking, posing a risk of corrosion. Where the water contains de-icing salt, brought into the car park on the tyres of cars, or in a coastal location, the corrosion risk is significantly increased.

18 Asset management

18.1 Introduction

Car park asset management is the process of safely maintaining the structure and its component parts predictably in service, over a known design service life, to deliver reliable financial returns. Good asset management, in engineering terms, is a whole life cost-balancing exercise of operational maintenance costs, capital replacement costs of items at the end of their planned life, and revenue interruption when maintenance works are required. It should be applied to all forms of asset.

Planned inspection, maintenance and replacement of routine M&E items in car parks is standard practice and is expected by the operator, much of it embedded in legislation, such as the Health and Safety At Work Act[68] and the Electrical Regulations[179] in the UK. Asset management systems, supported by the latest BIM records, allow these assets to be identified, their expected life determined, and preventive maintenance carried out before significant failure occurs.

The need to inspect, maintain and repair a car park structural frame over its design service life is a relatively new concept, with the need arising out of the sudden failure and collapse of several car park structures in the UK (e.g., the partial collapse of Pipers Row car park[180]) and elsewhere. In the UK, it has traditionally been assumed that reinforced concrete buildings constructed to current structural design and durability codes would not require significant inspection or maintenance over their design service life (typically 50 years).

By contrast, from the late 1980s, highway bridges and associated structures were found to be failing much earlier than the virtually "indefinite" life intended, with codes suggesting that 120 years was the minimum expected life.

Highway structures in the UK are known to be exposed to de-icing salts, and suffer near constant cyclic heavy loads, and since the early 1990s have been recognised as needing regular inspection, maintenance and where necessary, structural repair and strengthening works to maintain them in a safe and serviceable condition after only 20 years in service.

Car parks are more than just a building, and are more like highway bridges. They are commonly external and damp, and are exposed to the elements, including sea salt if near the coast. They take considerable vehicle loading when full, and receive de-icing salt run-off from contaminated snow on vehicle tyres and wheel arches.

As a result, car parks can begin to deteriorate structurally very early in their service life, principally due to undetected reinforcement corrosion. In addition to chloride ion contamination, structural deterioration can be caused by other mechanisms during service, including:

- Impact damage from vehicles
- Carbonation
- Alkali-silica reaction[109]
- Freeze/thaw
- Thermal cracking/movement
- Surface abrasion and wear
- Failure of protective coatings and waterproof membranes
- Poor quality or inadequate cover to steel reinforcement

Every owner/operator has a legal obligation to maintain their car park/s in a safe condition. This obligation applies to all persons entering the building, including the operator's employees, and includes those entering lawfully or otherwise. There are no exceptions. This requirement also extends to the immediate perimeter of the building to protect persons from falling concrete, or from vehicles which may accidentally breach the edge protection barriers or dislodge the cladding. Saving on maintenance does not always save money. Neglecting basic preventative care increases the likelihood of premature failure, the consequences of which can be significant.

Largely as a result of the partial collapse of the Pipers Row car park in 1997[180], the Institution of Civil Engineers (ICE) published new recommendations[181]. The purpose was to help inform owners and operators about the risks to the safe operation of their car park/s, and to explain best practice in asset management over the life of the asset. This requires appointing a suitably qualified engineer to provide a report that establishes the benchmark condition of the car park, including its structural capacity relative to current loadings, and any deterioration of the structural, drainage and waterproofing condition since construction, which is used to inform future inspection and maintenance activities.

ICE terms this report a car park 'life-care plan' (LCP), defined as:

> *"A documented managed approach to the inspection, maintenance and management of a car park structure, cladding, edge protection, drainage and waterproofing which is derived by the engineer, agreed with the owner/ operator and implemented throughout the entire service life of the facility, including demolition"*[181].

The difference between asset management and life-care is that asset management is the process for managing the facility to achieve the required financial returns[182], whereas life-care is the process for inspecting, testing and maintaining the car park to ensure that the required level of structural safety is maintained over the design service life. Effective car park asset management is therefore the combination of life-care and conventional fiscal management of the non-structural items of the asset, optimised to give best value.

With a new-build car park, the designer, contractor, owner and operator have an ideal opportunity to prepare the LCP, by recording the as-built design and construction of the asset, and then pre-plan future inspection and maintenance activities to ensure the continued safe operation of the car park. The following sections explain the approach needed to prepare the initial LCP at new-build. Further guidance on lifecycle operation of the LCP can be found in the Institution of Civil Engineers recommendations[181], and are not repeated here.

18.2 Car park deterioration and maintenance

However well designed or constructed, all structures begin to deteriorate as soon as they go into service. The design for durability (Section 11.10) allows the designer to select the materials and their corrosion protection systems to ensure that significant deterioration does not occur in the structural frame over the design service life.

Car parks may also inadvertently incorporate poor details that encourage deterioration, or errors in the construction and supervision, which result in parts of elements that are not as durable as intended. Local defects, varying from failed joints, unplanned and unsealed cracking, poor concrete compaction that causes honeycombing, and damaged/ defective membranes, can all contribute to reinforcement corrosion occurring within the intended 50-year design service life.

If the local defects are not recorded at the construction stage, or are ignored when they develop subsequently, deterioration will reduce the value of the car park and — if certain areas need to be taken out of service — affect the operator's ability to generate income.

The deterioration processes are complex and largely dependent on two inter-related variables; time and expenditure[183]. Once started, deterioration cannot be easily halted and the damage cannot be reversed, but it can be managed through a programme of structured and planned maintenance, following established procedures[129]. The warning signs are often ignored, possibly because repair and maintenance are considered operational costs, and are not properly assigned as planned capital costs. Therefore, undertaking unplanned operational maintenance will have a direct impact on profitability.

The answer is to ensure maintenance and repair works are predicted at the new-build stage, allowing the costs to be properly budgeted over the whole life of the asset. The LCP delivers the pre-planning necessary for proper whole life costing of future maintenance and repair.

18.3 Life-care plan at new-build

The LCP should contain all that is known about a car park structure. The structure is defined as the structural frame, but also includes the edge protection and perimeter cladding, which may be subject to accidental vehicle loads over the service life. For existing structures, the ICE recommends that the owner/operator should appoint an experienced chartered civil or structural engineer to be responsible for preparing the LCP, advising on all aspects of material durability, structural safety and the maintenance/repair strategy for the car park[181].

At the new-build stage if possible, it is most efficient for the original design team to prepare the LCP, as they will be best placed to advise on the particular features of the car park, based on their knowledge and supervision of the construction work. For new car parks, the LCP is usually prepared in parallel with the project handover information. The starting point is therefore the handover information pack (HIP) (Section 18.3.1).

The LCP is a living document, which must be updated as and when information changes. For new buildings it may comprise a single summary document, cross-referencing to other documents in the HIP. For an existing car park, it may comprise multiple files and reports depending on the age, complexity and condition of the car park.

By its nature, the LCP is ideally suited to being incorporated into an asset management database, allowing the information to stored, retrieved and updated easily, and which can be set up to automatically produce the required planned inspection and maintenance schedules, works proformas and procurement module and allow expenditure to be monitored. It can also be integrated with the electrical, mechanical, payment and security systems.

18.3.1 Handover information pack

The handover information pack (HIP) for all new structures should include a 'health and safety file'. In the UK, provision of a health and safety file is a mandatory requirement. It should identify the materials used in the car park and include reference to any unusual structural, mechanical and electrical facilities and identify operation and maintenance requirements, in particular:

- Description and general history
- Specifications
- As-built drawings
- Detailed records of construction and installation
- Details of warranties
- Any significant repairs carried out prior to handover, which may be a point of weakness in the future
- The presence and dangers of confined spaces (e.g., in lift machinery pits)

Notwithstanding the apparently open and ventilated nature of most car parks, accumulation of fuel fumes and exhaust gases is a significant risk to operations staff and users. In enclosed car parks (e.g., basements) adequate and well-maintained ventilation equipment is essential.

18.3.2 Initial life-care plan

The HIP forms the baseline information for the Initial LCP. Unless set up in a comprehensive BIM model, the HIP may not to contain all relevant information needed by an engineer when deciding whether the car park has been built to meet the design intent. Additional work may be required to fill any gaps in the information, including a review of records, visual inspection or limited testing if the quality of the construction is particularly poor. Examples of the types of information that need to be incorporated in the Initial LCP include:

- Design and construction features that may require special attention or inspection, such as bearings or cladding fixity
- Any location where the construction deviates from the design or specification, which may become a point of weakness in the future, such as low cover to reinforcement or poorly compacted/weaker concrete
- Construction-stage repair or strengthening to any damaged or otherwise defective areas, which may fail in the future
- Enhanced protection applied to the structural members, including deck membranes, anti-carbonation paints, deck joint systems and sealants

The Initial LCP should help owners avoid premature and unexpected damage that requires repair outside of the planned maintenance schedule. This should enable sufficient funds and resources, financial or otherwise, to be allocated at the outset, to maintain adequate levels of safety and to avoid property values diminishing. This is usually capitalised over either the design service life, or the operational life of an operator's concession (if shorter), on a whole life cost basis.

The Initial LCP should contain a programme for regular inspection and routine maintenance, and typical inspection levels are shown in Table 18.1[181].

Surveillance or routine inspections can only identify surface defects and cannot determine its structural adequacy or potential rate of deterioration.

Table 18.1: Types of car park inspection and requirements

Type	Action and frequency
Level 1	**Daily surveillance inspection** by car park operations staff, to identify obvious signs of damage (accidental or otherwise).
Level 2	**Six-month routine inspection** by trained inspectors working under the supervision of the engineer, recording any changes from the as-built information in the LCP.
Level 3	**Condition inspection** normally carried out when the structure is six years old and repeated at the intervals recommended in the LCP. The purpose is to identify, through testing and inspection, any premature deterioration of the car park, with the first such inspection termed the **baseline condition survey** (Section 18.3.3). Set inspection intervals using a risk-based assessment and included in the LCP. If the design and construction of the car park has no known significant deficiency, the repeat period can be 12 years.
Special	**Special inspection** is a targeted investigation into the cause and extent of unexpected deterioration (e.g., ponding, cracking, seepage, joint leakage, deterioration in waterproofing membranes, etc.), identified through the Level 2 or 3 inspections.

Structures designed to the recommendations in this guidance should be less likely to suffer premature deterioration than earlier car park designs. However, when a routine or condition inspection indicates there is unexpected deterioration, the frequency and types of inspection will need to be reviewed, and may trigger a Level 3 and/or a Special inspection.

The output from all inspections should be recorded, using drawings, logs, photographic records and test results where appropriate, identifying to the engineer any items which require immediate attention or further investigation. Following the Level 3 or Special inspections, the report should identify any deficiencies in the construction and provide recommended actions. These could include any immediate actions and/or whether works should be scheduled in the future, such as further surveillance, inspection or repair (Section 18.3.4).

Warranties, completion certificates, product data sheets, health and safety requirements, etc., should also be filed in the LCP. Above all, the LCP should be simple and easily understandable by the intended users of the document, including the owner/operator, staff carrying out inspections and the engineer. Ideally, the documentation should be kept by a registered keeper who has an understanding and appreciation of its purpose and importance[181].

As well as inspection intervals, the LCP should provide a programme for all routine maintenance and capital replacement works to be carried out at manufacturers' scheduled intervals to maintain the structural performance of the car park, including:

- Mastic joint sealants, typically with a life of 10 years, which fail by splitting and debonding
- Deck waterproofing, typically with a life of 10–20 years, which fail by splitting, debonding or wear
- Anti-carbonation paint systems, typically with a life of 10–20 years, which may become ineffective over time
- Special ancillary materials, such as bearings, movement joints, cladding fixings, edge protection systems etc., which will have a finite life
- Any concrete repair and strengthening works, including corrosion control systems such as impressed current cathodic protection

A typical schedule of car park maintenance actions is given in Table 18.2.

Table 18.2: Car park maintenance schedule: best practice

Item	Interval
Benchmark condition survey	At construction
Inspections	
Routine maintenance	Monthly
Drain cleaning	Every six months
Wash-down and deck cleaning	Every six months
Cosmetic repairs	As necessary
Surface protection	Every 10–15 years
Preventive repairs	As necessary
Structural re-appraisal	Every 16 years
Structural repairs	As necessary
Refurbishment	Target date
Demolition and replacement	Design service life (unless extended)
Design service life	Set by stakeholder
Works value	Annual base cost
Whole life value	As necessary
Current value (if defect free)	As necessary
Residual value after works	As necessary

18.3.2.1 M&E plant

The LCP maintenance and capital replacement programme can also contain operation and maintenance requirements for the mechanical and electrical plant. Previous chapters in this guidance provide detail on the types of equipment used for building services, traffic management (including vehicle access controls and signing), on-site staff accommodation and fire detection and prevention systems.

18.3.3 Condition survey and structural appraisal

For new car parks, a benchmark condition survey is not usually required, unless the structure is known to have an issue during construction. The benchmark condition survey can then be carried out, documented, and included in the Initial LCP.

For older structures without any prior inspections, the ICE recommendations[181] give the following advice (also refer to Table 18.1):

- **Benchmark condition survey:** for structures up to six years old, establish the level of deterioration, followed by further (Level 3) condition inspections and special inspections as required
- **First full structural appraisal:** for structures more than 12 years old that have no record of an appraisal being carried out, this should be undertaken without delay. If the LCP is also missing, the first point of reference should be the health and safety file, which will make the task of identifying materials used in the car park much easier, and will usually include reference to any unusual structural, mechanical and electrical facilities and maintenance requirements
- **Repeat structural appraisal:** where necessary the LCP should recommend future structural appraisals to evaluate the integrity and adequacy of the structure, edge protection and cladding. The appraisal should be supplemented by special inspections as necessary. The scope should include consequences from any changes in code provisions, vehicle loadings (e.g., increased electric vehicle use) and changed car usage over the various levels

Some of these recommendations are directly relevant to the refurbishment, upgrading or extending of an existing car park (Section 18.4).

18.3.4 Maintenance options

The owner/operator, working in conjunction with the engineer, can develop a range of different options for each action, taking into account parameters such as capital cost, life expectancy and client cash flow preferences, as well as considering the whole of life cost-benefit, rather than just the maintenance expenditure[181].

Figure 18.1 illustrates that if premature asset deterioration occurs and is left unchecked, it can result in critical deterioration that poses a threat to the stability of the car park. Two contrasting approaches are shown to restore the functionality of a car park:

- Substantial expenditures (Repairs 1 and 2), intended to restore the car park to a condition close to its projected ideal life curve
- Lower cost, regular maintenance actions to keep the car park above a critical limit

Figure 18.1: Possible asset deterioration profiles and maintenance actions

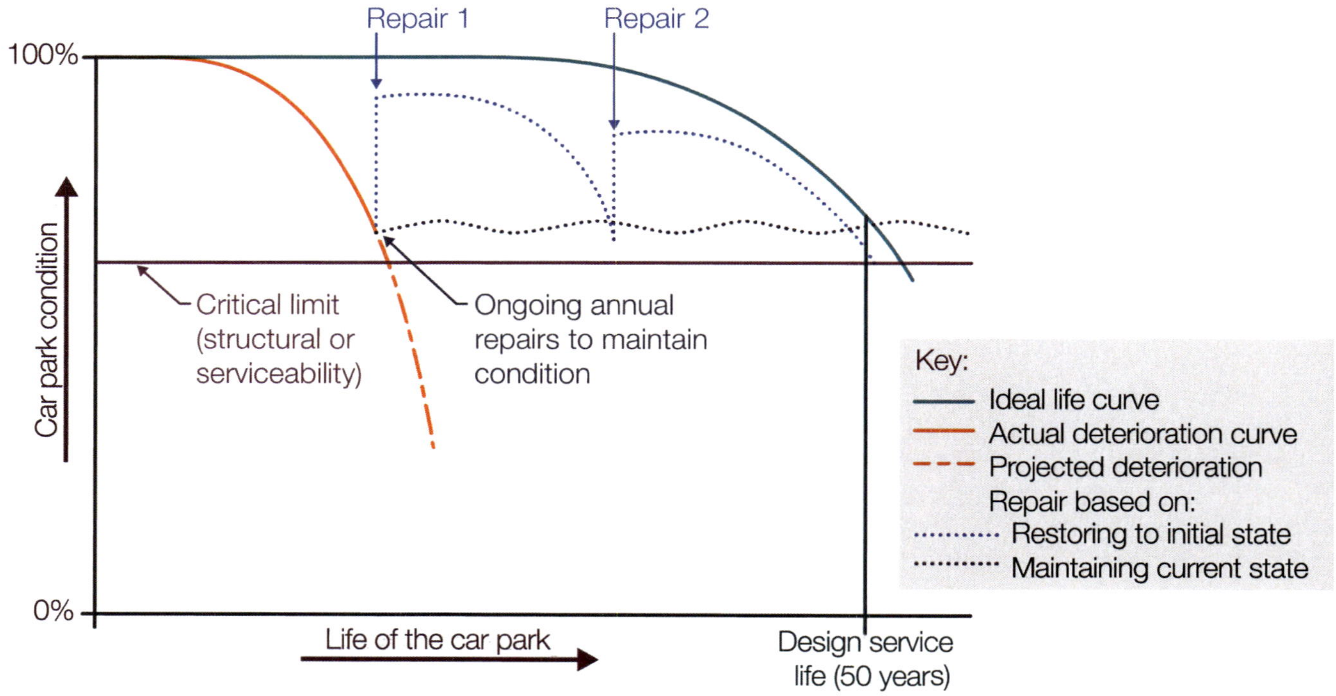

18.3.5 Development of the LCP

The LCP is a living document. It must be continually updated and revised based on the latest information.

Feedback from all inspections (Section 18.3.2), surveys and structural appraisals (Section 18.3.3) and maintenance actions taken (Section 18.3.4), along with any changes or capital expenditure to replace or strengthen deteriorated elements (Section 18.4), must be immediately captured and updated in the Initial LCP.

The documentation is likely to become extensive, particularly after reaching the structural appraisal stage, and is best suited to a database recording system where there is more than one car park in the portfolio.

A summary schedule should be produced in chronological order with future actions indicated at key stages.

18.4 Upgrading existing structures

The ICE recommendations[181] give guidance for existing car parks. The approach will be relevant to new-build designers in situations where an existing car park is being extended, as well as to the engineer maintaining an existing car park.

The first step before considering the detail of the upgrade is to ensure the LCP is in place. Where it has not been updated, or is absent completely, a retrospective LCP should be produced that considers among other things:

- Structural adequacy of the car park in the as-built condition
- Any changes in design codes and vehicle loading since the time of construction
- Effect of deterioration in concrete, steel or other materials (e.g., previous repairs and strengthening) that could influence the structural performance

The structure will have been designed for a specific loading and maintenance regime, which can include some provision for changes in use or modifications. The LCP should make clear any particular limitations that apply to:

- Extensions and increased height
- Loading for alternative uses and during works on the structure
- Surfacing thickness (affecting car headroom)
- Planting and landscaping loads
- Drilling through elements and opening up (e.g., sensitivity of structure near columns in flat slabs)
- Areas not designed for chloride resistance, where salt should not be applied for de-icing
- Washing of vehicles, which should only be carried out in specific designated areas and where a waterproof membrane has been applied
- Cleaning of the decks, ramps and drains, including a regular washing-down at least annually at the end of winter to reduce the build-up of de-icing salts — ideally by wet vacuum to remove surplus water
- Performance of fixings and edge barriers
- Zoned areas designated for fuel storage

As noted in Section 11.10, with good drainage and airflow, reinforced concrete car parks are tolerant to relatively high levels of chloride ion build-up in the concrete surface. Changes in use of particular levels of the car park may significantly affect the durability of the structure. Actions that lead to increases in humidity or physical wetting may accelerate the deterioration processes, by providing moisture to fuel the corrosion process[129].

For more detailed information on upgrading existing structures, refer to Chapter 19.

19 Upgrading and modifying existing structures

19.1 Introduction

As discussed in Chapter 10, the decision to futureproof the design of a new-build MSCP is a challenging issue with many factors to consider. While this is a consideration for virtually every client at the design stage, it remains relatively rare for new MSCPs to be designed for alternative future uses. Instead, priority is generally placed on lean design, being as efficient structurally as possible, and targeting minimal embodied energy in the purpose-specific car park building. This is in line with the hierarchy of net zero design[45], where the designer should first seek to reduce the materials used in the construction.

This chapter reviews the principle and feasibility of upgrading and modifying existing car park structures.

19.2 Upgrading (extending existing use)

An MSCP is a relatively simple structure consisting of a steel or concrete frame, decks comprising precast concrete or composite (steel and *in situ* concrete) sometimes with a screed, deck coatings and other components including lifts, lights, a fire alarm system, PV panels, battery rooms etc. Occasionally the MSCP might include some other spaces such as staff welfare or office space, WCs and small commercial units. In general, older MSCPs (built during the 1960s and 1970s) still in use are typically constructed with concrete frames, whereas the trend in recent years has been towards steel frames, reflective of advances in material technologies and market trends in material costs. Cladding systems have also become increasingly sophisticated and elaborate.

These structural and building components may be well suited for repurposing, or may already be past their design service life or in a poor state of repair and more suited to recovery or recycling.

While there is an overriding need to ensure that any construction works have the minimum negative impact on the environment, buildings — including car parks — do become redundant; often as a result of societal change. Even though they may not be life expired in terms of condition, renewal under these circumstances may be the only viable option.

When assessing the potential of an existing car park facility, consideration may be given to continuing its use as a parking facility or changing its function.

When considering refurbishment in order to extend service life, the design team will need to consider a range of factors:

- The MSCP's state of repair and its previous maintenance history. Is it in good condition, architecturally and structurally? Generally, this review will need to be undertaken by both architectural and engineering consultants, and will take into account both finishes/aesthetic properties and the underlying structural condition. Some aspects of structural degradation may be easily apparent to the layperson (for example observing cracks, exposed concrete rebar or fallen debris) but others may not be immediately visible (such as moisture content, honeycombing or missing reinforcement). A range of non-intrusive and intrusive tests may be required, with results presented in a structural appraisal report
- If the MSCP does require maintenance or repair and rectification works, the structural appraisal should advise on the urgency of relative tasks and the implications of not undertaking the works. The client may then review the financial impact of the recommendations (including direct costs as well as the lost revenue/disruption during the works) and decide whether to proceed
- As part of the structural appraisal, the building owner should make available any maintenance and repair records. The future maintenance requirements, once the refurbishment works are completed, should also be programmed and costed so that they can be factored into the viability assessment, and the building owner made aware of their obligations

- The design team will also need to consider the original design specification, in particular the structural loading and the natural frequency. These should be established from the building manual or life-care plan (LCP), or from investigative work where records are not available. They can then be compared against the contemporary requirements and any shortfalls identified. The design team can then recommend how to address any shortfalls, and a cost plan can be prepared for any works required. It is highly likely that the residual strength of an existing MSCP will be lower than when it was first constructed due structural deterioration while in service, and therefore its future capacity or suitability for contemporary vehicles could be limited unless works can be undertaken to ensure that it has sufficient structural capacity. Clients also often wish to test the feasibility of adding floors to existing MSCP structures to increase capacity. Similar studies will confirm whether this is feasible
- Inside the MSCP, especially in older structures, there may well be interbin columns. These can present constraints on re-lining or amending the layout to modify the parking distribution and circulation e.g., if the client wishes to increase space sizes to suit modern larger vehicles. This may restrict the overall capacity of the MSCP and impact on the financial viability of the refurbishment
- The design team will review the existing building against current regulations and highlight any deviation from contemporary standards. This can result in substantial modification e.g., the introduction of an additional fire escape core
- Older MSCPs are more likely to employ older circulation systems such as scissor ramps or split-deck arrangements (Section 5.4). These may have restricted dynamic capacity compared to more contemporary arrangements, especially with the increased size of vehicles. The ease of navigation may also be compromised compared to contemporary user expectations. Unless addressed, this can result in damage to vehicles (from collisions with the structure and other vehicles) and potential congestion at peak times. It is possible to carry out major alterations to improve an existing MSCP (e.g., changing from narrow scissor ramps or split-level decks to a three-ramp arrangement), but cost and disruption can be substantial and therefore viable only in limited circumstances
- As part of any refurbishment works, the car park owner/operator is likely to wish to improve the parking technology and management strategy. This could include the introduction of ANPR cameras and paperless technology, for example. The design team will need to review the existing provision of power and data throughout the building and identify the feasibility of introducing new routes where necessary, which could involve the perforation of existing structural members or fixing new ducting. The new equipment (e.g., ticket machines, barriers, cameras etc.) will also need to be identified, and sufficient space allocated within the floor plan
- The addition of EV charging facilities is also a common aspect of many refurbishment schemes. In order to do this, the design team will need to establish if the existing electrical supply has capacity for the projected demand (which will vary depending on the number of spaces, the charging rate and the diversity requirement) or if a new substation may need to be provided. This will require discussions with the local distribution network operator (DNO). If it is established that there is sufficient capacity or it can be provided, the design team will need to identify the location of the EV spaces within the car park and design the power and data routes required to serve them. This may include new risers and plant rooms as well as the charging equipment itself (floor mounted pedestals or wall mounted chargers)

The relevant party (whether building owner, leaseholder or operator) will need to carry out a financial viability assessment on the refurbishment works in order to decide whether to proceed. This will consist of three components: the cost of the alteration works, the cost of future maintenance, and a comparison to the cost of an equivalent new-build car park. The new-build scenario may also offer scope to increase the height or footprint and therefore capacity of the car park, while also providing an improved user experience and opportunity to fully meet contemporary standards. This would also increase the service life and revenue generated by the facility.

The owner/operator should consider their responsibility to act in a sustainable manner by fully considering the embodied carbon/energy impact of the proposals. A new-build facility might have a much improved operational energy demand due to more efficient fittings etc., but this may not be outweighed by the demolition and new-build carbon impact. A 2011 report by the US National Trust for Historic Preservation found that:

> *"it takes between 10 and 80 years for a new building that is 30 per cent more efficient than an average-performing existing building to overcome, through efficient operations, the negative climate change impacts related to the construction process."*[184]

Specialist consultants can undertake detailed carbon calculations to inform this decision and to provide justification in planning policy terms where required.

Finally, the design team should review the relevant local planning policy — as certain adopted policies may support the retention, alteration or even removal of an existing car park facility (for example where a 'low-value' building is located within a sensitive conservation area). If the proposals are not policy compliant, they may be entirely futile. Engaging in pre-application enquiry with the local planning authority can give comfort on this at an early stage.

19.3 Modifying (enabling alternative use)

Alternatively, the car park facility may no longer be required and the owner may wish to explore repurposing the building. This might be because demand has fallen to unsustainable levels — perhaps the car park served a destination which no longer exists, or because an alternative facility offering an improved user experience has become available locally.

Current UK Government initiatives, both national and local, promote greener modes of transport and healthier living by encouraging "last-mile" travel and goods distribution on foot or by cycle. In this scenario, people park their cars at the edge of a district or settlement before swapping to an active travel mode for the final part of the journey. This shift might result in existing car park structures at the edge of towns becoming more desirable, providing an opportunity to bring structures back into use or providing an incentive for owners to update the facility.

The anticipated gradual increase in the proportion of autonomous vehicles on the national highway network will have an impact on parking structures generally (Chapter 10). However, in terms of its impact on the repurposing of structures, the implications around structural loading, charging requirements and fire strategy (amongst other topics) can be predicted to be more onerous than those for non-autonomous vehicles. As when designing a new-build car park, these factors should be considered early in the design process.

There are two separate tasks in order to establish what alternative functions might be feasible. The first is typically undertaken by the client's own team (typically a land agent) and will be to establish what demand exists in which sectors, and what the market expectations are for that type of facility. The second is undertaken by the design team and initially, consists of similar studies to those required for a refurbishment project (Section 19.2).

The design team will therefore need to establish what the structural capacity of the MSCP is, including structural loading and natural frequency, and to review how this might be increased if required for the new uses under consideration. Most existing car park structures were only designed to provide 2.5kN/m^2, which is inadequate for most repurposing functions. If the existing structure can only be upgraded to a certain point, as is likely, this will limit the range of future uses. This could be constrained further by the ground conditions e.g., if the primary frame can be upgraded or reinforced but the ground can only accommodate a certain load before new or additional piling is required.

The floor-to-floor height of the existing car park is also likely to present another major constraint. Most other multi-storey building types such as offices, housing and healthcare, require a significantly greater floor-to-floor dimension than a car park in order to accommodate services, equipment and provide a quality living or working environment (Case study 1, Chapter 8). Older car parks also typically have a lower floor-to-floor than contemporary MSCPs and can be as low as 1.9m clear height, which is suitable for very few alternative uses. In some exceptional circumstances and with sophisticated structural engineering, it may be possible to remove every other floor to increase the height of each new level to twice the existing.

Ramped or sloping car park decks can also present a challenge for conversion, even where this is only a minor gradient for surface water drainage e.g., 1:60. Most habitable uses will not tolerate a sloping floor (e.g., workspace or housing) and therefore provision needs to be made to install a second floor above the existing deck to provide a level surface. This further compromises the potential clear height.

The depth of plan of the existing MSCP can also represent a design challenge for many alternative uses, as the amount of natural light and ventilation reaching the centre of the building can be limited. This would be particularly inappropriate for residential use. A solution may be to open up the centre of the building and create an enclosed atrium or open courtyard, but this will entail major structural works.

Some existing car parks may feature a full or partial roof on the top deck. This may be appropriate for retention or may need replacing, depending on the proposed use and its existing properties. If it is to be retained, consideration needs to be given to maintenance access and to how the structure might be upgraded for thermal and acoustic insulation requirements.

As part of most repurposing exercises, it will be necessary to add additional services and service runs to the building. As well as confirming the capacity of the existing incoming supplies and seeking extra capacity if needed (e.g., for electricity and drainage), this may include brand new connections. These will need to be discussed with the relevant district network operators/utility suppliers. Once the incoming supplies and outgoing drainage/sewerage connections are confirmed, the route of services within the building needs to be carefully planned to serve all the units or outlets required for the new function. For a conversion to an inhabited function such as housing, this could include multiple sewerage runs and domestic cold water supplies, which would occupy a significant area both vertically between floors and in plan.

The use of domestic gas boilers in UK housing is being phased out in favour of more sustainable technologies such as air source heat pumps. The developer may wish to consider the provision of network heating and cooling systems for the building rather than providing separate systems for each unit, whether these are commercial or residential, particularly where they intend to retain the freehold interest in the building. Furthermore, most car park structures are not designed for disproportionate collapse resulting from a gas explosion and therefore any form of heating that relies on burning natural gas and/or hydrogen would not be possible.

Existing car park structures are naturally unheated and uninsulated, and often naturally ventilated without any mechanical equipment. Many alternative uses require thermally insulated and heated spaces; the provision for which must be factored into the design for a conversion. Cold bridging can be a particular issue given the typical car park structure with exposed primary frame elements, whether steel or concrete. Overcoming this will typically require overcladding of the existing structure, which can drastically change its appearance. Insulating spaces within the car park can also significantly reduce the available floor area and internal clear height for the alternative use.

The developer may wish to obtain a recognised sustainability accreditation for the building such as BREEAM[58]. This provides reassurance to occupiers that the building meets a certain standard and may also be expedient for the developer for their sustainability objectives as a business. The Building Research Establishment (BRE) run a framework specifically for refurbishment and fit-out, which is likely to be the most appropriate route to accreditation. If this is anticipated to be a requirement of the project, advice should be sought from a specialist consultant at an early stage as targets must be identified and achieved from the very beginning of the design.

With any alternative use, the fire safety strategy will change and will often become more onerous than that of the existing car park use. The support of an experienced fire engineer will be essential to confirm the new strategy. It is likely to have major design implications and could impact anything from the structural performance to the escape distances and number of escape cores or compartmentation of the building.

Most alternative uses require a greater fire resistance than an open-sided car park and therefore it will be necessary to upgrade the performance of the frame, particularly where the existing building has an exposed steel frame. It is currently theoretically possible to upgrade a steel frame from a 15-minute rating to an enhanced rating using intumescent coatings which can be retrospectively applied, or by covering exposed steels with fire-rated boxing or encasement in concrete. The fire engineer will be able to advise on the actual fire rating required, as well as firestopping requirements (e.g., to stop the spread of flame between floors or compartments). It will also be necessary to install a new fire alarm system and give consideration to fire suppression, which may be mandatory for some alternative uses (and could have significant consequences, such as the provision of water tanks for sprinklers, for example). These actions have both spatial and cost implications that should be taken into account in the feasibility assessment.

The overall viability assessment will help identify what sort of uses may be feasible based on the previously-mentioned factors and the financial model for the development. In general, the lower the intensity of the proposed new use, the easier it will be to repurpose the existing facility. A range of case studies for the repurposing of MSCP structures exist in countries around the world, including:

- **Urban farming and hydroponics.** These can make use of the exposed top deck of the car park for container planting in sunlight, while lower or basement floors can be used for mushroom farming or hydroponic farming with the use of artificial lights. Hydroponic farming has a particular advantage in repurposing situations, as the growing medium (water) is of lesser structural load than soil-based farming, which can add huge structural loads. Farming uses do not usually require insulation or space heating, and conversion can be relatively straightforward. The location in city centres with good transport links also allows for effective distribution to market. Case studies include La Caverne in Paris and rooftop farms in Singapore

- **Leisure functions such as open-air cinemas, rooftop bars and spaces for exhibitions.** As with farming, there is little need for space heating or thermal upgrade to the structure. However, with greater occupancy rates than the original car park usage, there can be a structural live load implication and a need for additional fire escapes. A very well known example in the UK is the Peckham Levels building in East London. It provides these amenity functions alongside start-up business units, office space and shared working spaces
- **Relatively informal business premises e.g., business incubators or start-up spaces.** These users may tolerate or welcome a slightly 'edgier' setting than a large corporate operation and may be more flexible in terms of clear height and services if the rental rates are competitive. These uses will require additional services, especially heating (with associated thermal insulation), power and data connections. Peckham Levels incorporates a number of these units, as does the Buckingham Road MSCP in Worthing, West Sussex. They are typically located around the perimeter of the building in order to avail of natural light and ventilation
- **Residential.** This is relatively rare but a concrete framed split-level MSCP in Bolton was granted planning consent in June 2022 for conversion to 116 residential units and three commercial units[185]. The public car park function is no longer required or financially viable, so the scheme proposes to remove the central split-level ramps and replace this zone with an enclosed atrium. The residential units are effectively single-sided with one aspect to the perimeter of the building and one into the atrium. The floor-to-floor height is 3m which will have implications for the clear height in the units and it is not clear how the designers intend to deal with the drainage falls of the existing deck. The concrete frame is overclad with a brick skin with regular domestic-scale windows serving each unit, thus enabling the designers to improve the thermal performance of the building.

As with refurbishment for future car parking use, the design team should review the local planning policy context and assess the proposed uses for alignment with policy. This may be in terms of providing employment sites or meeting housing targets, or regenerating a town centre by providing leisure uses.

19.4 Disassembling/relocating

If repurposing is not of interest or is concluded to be non-viable, the building owner may like to consider the relocation of all or part of the existing structure. A number of car park systems have been marketed over the years as having the potential to be disassembled and reconstructed on an alternative site. In practice there are a range of constraints that could influence the feasibility of this strategy:

- Some structural systems are inherently more appropriate for disassembly than others; e.g., a Meccano®-style kit of parts single deck system may be unbolted with all components saved for reuse, whereas structures that have some elements of *in situ* concrete in decks or wearing screeds cannot be disassembled without causing significant damage, even though the primary frame elements can be salvaged for reuse
- The characteristics of the existing and new site may differ in terms of gradients, ground conditions and relationship with the highways network and pedestrian destination; meaning that the column lengths, bracing, ramps and stairs need to be reconfigured within the footprint of the building. This may or may not be structurally possible depending on the exact nature of the building
- The costs involved in disassembling the structure, transporting it and reconstructing it may be prohibitive. This work may need to be carried out by the original supplier in order to maintain any warranty, or by a specialist who is willing to take on the risk of this work (inevitably at a financial cost to the client)

Following the waste hierarchy[45], the conclusion may be that the existing car park cannot be refurbished, repurposed or reused. It therefore needs to be removed from the site and the maximum proportion of constituent parts and materials recycled.

Demolition firms will salvage and reclaim as much as possible where there is a financial incentive to do so. This is unlikely to offset the cost of the demolition to the client, to whom there will be a financial cost rather than any kind of payback. The market in recycled materials fluctuates and what is considered valuable one year may not be the next. In general, metals are consistently more valuable than concrete but the latter may still be crushed and reused for hardcore in future construction projects. Car parks increasingly contain valuable materials in the form of precious and heavy metals in batteries, PV panels and data systems etc. These are likely to be prioritised for recycling by any demolition contractor.

The demolition contractor will be responsible for ensuring that any noise, run-off and dust resulting from the works do not cause a pollution hazard or nuisance to the local community. A method statement will be required by the local authority prior to any works taking place. Dust is a particular hazard to health when demolishing concrete structures, and dust suppression is likely to be required. Control of vehicle movements is also tightly monitored and working hours can be expected to be pre-defined, in agreement with the local authority.

Whether or not it is explicitly set out in contracts, the car park owner has a social and ethical responsibility to interrogate their supply chain and ensure that all contractors have appropriate licences and approved disposal supply chains to avoid 'offshore' dumping of waste in other countries, or unlicensed disposal within the UK.

20 Special structures

20.1 Introduction

When visualising a parking structure, the immediate image is of a multi-level building where vehicles drive onto a deck, along an aisle and into a parking bay, either at 90° to the aisle or angled off the aisle. The focus of this guidance is how to design and build such facilities. However, there are a number of other types of car park structure and this chapter provides an introduction to some of the most common alternatives.

Mechanical or 'automatic' car parks have been around for almost 120 years, almost as long as conventional car parks. Commonly referred to as 'robotic' car parks today, these are essentially machines that receive the car from the driver at the site entrance and store it until required (Section 20.2). Drivers do not enter a robotic car park beyond the portal.

Another type of parking structure are buildings that are designed to accommodate block or 'echelon' parking (Section 20.3). These are facilities where again the car is collected from the driver at the entry point by a member of staff (valet), and parked line astern in rows. There are no aisles, other than at the front and rear of the deck. Sufficient space is left between rows of cars to allow the valet to enter and exit the car. The cars are parked more densely than in a self-parking car park, so the deck loading is increased.

What makes these types of car park different, is that for both robotic and block parking, the driver does not park the car in its final storage location. With a robotic car park, the driver leaves the car in a reception area. The car is collected by a mechanism that takes it away and stores it within the machine. With block parking, the car is also left in a reception bay by the driver before being taken by a valet to a collection area.

Stacker systems are becoming more common as a way of increasing capacity, albeit at the cost of less immediate access to the vehicle (Section 20.4). Finally, Paternoster car parks are described briefly (Section 20.5). This type of parking facility is extensively used in some countries (notably Japan) where land is at a premium, but they are not widely favoured in Europe.

20.2 Robotic car parks

Some of the earliest robotic car parks were built in the USA in the 1920s by Kent Automatic Garages. Cars were moved vertically in lifts and horizontally on the floors to and from the lifts by "electric automatic parkers" essentially small electric tugs. These early car parks were monumental structures, up to 25 storeys tall and capable of holding 1,000 cars; all built on a plot of just 30m x 43m. It seems that the business eventually failed as cars became bigger, reducing the effective capacity of the car parks. An early example in the Columbus Circle area of New York was eventually converted to office use and is currently still in use as student accommodation (Figure 20.1a and b).

Today, there are of two main types of robotic car park. The first is in many ways similar to the Kent model with a traditional multi-floor structure where vehicles are moved around on pallets (Figure 20.2). The driver will usually bring the car to an entry point resembling a single-car garage, where the car is parked on a pallet and locked. Usually, the car will be scanned to check its dimensions and that it can be accommodated by the facility. Once the car has been locked and left, it will be picked up by a skate under the car pallet that lifts the vehicle clear of the ground and moves it into the facility. The skate can move horizontally on a floor in two dimensions and the pallet, with the vehicle, can also be moved vertically between floors via a lift. Vehicles are stored on floors often in echelon ('angle') and can be moved and repositioned as required to recover other vehicles.

The second type has no floors as such. It is in effect an automated car warehouse with a framework housing a grid of storage 'bins' for cars. Cars are moved horizontally and vertically on a mechanism that can perhaps be best described as a forklift truck on rails. Having received the arriving vehicle on a pallet, the machine moves sideways on rails and up and down on a lift until the vehicle and pallet is inserted into a vacant bin (Figure 20.3).

Figure 20.1a: Early robotic car park by Kent Automatic Garages. Columbus Circle, New York, USA

Figure 20.1b: Columbus Circle garage building today: mixed-use office/residential space

Figure 20.2: Multi-floor structure

Figure 20.3: Framework of storage 'bins'

For both types, depending on the design and manufacturer, the pallet may be capable of rotating through 180° so that the returned vehicle is presented to the driver facing forward to be driven out. This is not necessary of course if the entry and exit points are on opposite sides of the structure.

The biggest advantage of a robotic car park is that it can be built on a smaller footprint than a traditional self-park car park, principally because there is no need to be able to accommodate lengthy ramps to move between floors and relatively wide drive aisles. This is particularly useful when parking is needed below buildings and the intrusion of the access ramp into the building floor plate would be unacceptable/impossible.

Cars in a robotic car park are isolated from people and vehicles. This largely eliminates the risk of the vehicle being broken into, or damage from collision with another car, or the structure, unless the machine malfunctions. Oversize cars will be rejected at the entrance. The storage bins/parking bays can be smaller than a standard parking space, as there is no need to allow for doors to be opened.

Lighting and ventilation requirements are much reduced compared to a conventional car park. Lighting provisions could feasibly be reduced to maintenance or security lighting levels and operationally are only required when service personnel are present. Some manufacturers suggest that no ventilation is required but this seems inadvisable in a structure containing vehicles that could be venting hydrocarbon fumes in the vicinity of electrical equipment that could produce sparks. However, as there no engines running within the structure there would be no exhaust fumes and with no pedestrians, the level of ventilation can be much reduced (potentially down to one or two air changes per hour, similar to a plant room environment). Note, that despite reduced ventilation requirements, there may remain a need for smoke extract systems as part of the building fire strategy.

There is no need for ramps, lifts, or staircases, other than for service access. Cars can be parked in echelon and aisles are fewer and narrower, meaning that a robotic car park will be more space efficient than a self-parking car park. The overall saving on height is less significant if the robotic car park is to house the standard design vehicle (Section 5.2.1).

Compared with a conventional car park, a robotic car park carries an increased load, imposed by the parking machinery, and this must be allowed for in any structural calculations.

Although a robotic car park will use little or no energy for lighting and ventilation, this has to be weighed against the considerable energy required to move the cars in and out of the facility. This activity also imposes a considerable maintenance requirement on the site operator, compared with a self-parking facility. Anecdotal data from the US suggests that the maintenance costs are about three times that of a standard car park.

Access to a robotic car park is usually via a series of entry portals where the driver parks and the machine then takes charge of the vehicle. These perhaps present the most significant disadvantage of the robotic car park when compared with a conventional facility. Manufacturers typically claim that their machine will park/unpark a car in 1–2 minutes (or 30 per hour) per portal. By comparison a rising arm barrier/ticket reader can pass more than 300 cars per hour. Note, these numbers are based on very idealised conditions, making no allowance for extra activities that a driver and passengers might undertake when entering/exiting a car.

A common concern with robotic car parks is: What happens when the machinery malfunctions? Certainly, some early systems suffered from chronic unreliability with constant breakdowns and vehicles becoming stuck. Modern facilities are much more reliable, with aircraft-style multiple redundancies built into critical systems. Unfortunately, like aircraft, if there is a failure the results can be very serious, with vehicles trapped inside for hours or even days.

20.3 Echelon, valet or block parking

'Echelon', 'valet' or 'block' parking refers to a facility where cars are parked close together to maximise the number of vehicles that can be stored in a given space. For the purposes of this guidance the term 'block parking' will be used.

The concept is quite simple. Cars are collected at the entrance by a valet, who will drive the car to the deck and park it as close as possible to the car in front in line astern. When the driver is ready to collect the car, a valet recovers it to a holding area, so that when the driver returns the car is ready to be driven away.

The term 'valet parking' is widely used to describe parking at places like hotels and restaurants where customers' cars are collected from them when they arrive and parked nearby. Usually though, this is in a surface lot or a convenient self-park car park. The term is used here for places like airports, where cars are collected at the terminal and driven to a remote location for storage, either on a surface lot or in a purpose-built structure (Figure 20.4).

Figure 20.4: Block parking: Gatwick Airport, London, UK

As cars are stored in blocks it is difficult to recover a vehicle mid-block if the parking was random. To operate efficiently, cars are organised into lines of vehicles that are planned to be recovered at the same time. This is possible because usually, the people leaving their cars would be returning at a known time, on a known flight. In this respect a block parking car park operates rather like a railway marshalling yard where arriving cars enter from the back of the car park and are lined up in "trains" of cars that will leave the parking area at the same time.

If a car park is designed for block parking, the floor loading will be greater than for a conventional self-parking car park and the designer will need to make an assessment based upon the preferred configuration of parking areas and drive aisles.

A block parking car park will not need the same levels of lighting as a conventional car park. Lighting is required for safe manoeuvring and staff safety only. Similarly, stairs are required for staff movement and fire safety only.

Parking cars in close proximity increases the risk of fire spread beyond the original vehicle, and so consideration needs to be given to limiting the risk of spread, or the maximum size of conflagration. Thought should also be given to ensuring adequate access for firefighting operations.

20.4 Stacker parking systems

Stacker parking systems aren't really parking structures per se; rather, they are mechanisms that can be installed within parking structures to multiply the capacity of the car park by parking one (or more) cars above one another. Modern systems now include EV charging points.

The most basic system places one car above another on a platform that can be lifted above the ground to allow a second car to be stored under the first (Figure 20.5). Although this doubles the capacity of the car park, it reduces utility for drivers, as the higher car cannot be accessed if the ground level space is occupied.

There are more advanced versions available where two or more cars can be parked and accessed at will (Figure 20.6). To achieve this there has to be a below-ground pit to allow the whole parking stacker to move vertically, like a lift. If the lower parking space is to be accessed, the higher spaces have to be lifted above ground level to allow a vehicle to enter/leave the lower space. Once this vehicle has parked; it can be lowered into the ground and another vehicle parked. Both vehicles can still be accessed at will.

Figure 20.5: Basic stacker parking system

Figure 20.6: Advanced stacker parking system

A requirement of this type of system is the need for space below the regular floor level and, to accommodate the device, the parking deck would need at least double the standard headroom.

Stackers where the upper car is on an inclined platform require a greater headroom to accommodate them but not as much as twice the headroom as the cars are stacked similar to those on a vehicle transporter. Another drawback of this system, in addition to the lack of easy access to the vehicles, is the potential increase in intensity should one of the vehicles catch fire. This would increase the fire load on the soffit of the structure owing to the proximity of the upper vehicle.

Where plot size is limited, a car stacker system can be installed underneath a driveway, for example, to enable cars to be parked in the ground by mechanically raising a parking pad in the driveway to allow the second car to drive under the pad and be lowered into the ground. It is possible to park more than one vehicle in the ground by using a platform to move the vehicles horizontally. This type of installation may be accompanied by a mechanical turntable at the rear of the driveway to enable the vehicles to be turned through 180°.

A limitation with car stacker systems is that they are not really suitable for casual visitor use, as they require the user to be familiar with the operating system.

20.5 Paternoster car parks

The paternoster or 'ferris wheel' car park has been in use for at least a century. The mechanism has a continuous belt that moves vertically. Cars park on pallets that are hung off the belt that remain horizontal as they are taken up into the air as vehicles are parked, and then brought down again as required.

The design is very space efficient as it only requires a footprint slightly larger than two parking spaces and can park many vehicles. The actual number is dependent on the tower height but structures with capacities as high as 48 vehicles have been reported. These structures remain popular in Japan, where land is at a premium, but less so in Europe and North America.

References

1 The Institution of Structural Engineers. *Design recommendations for multi-storey and underground car parks.* London: IStructE Ltd, 1976.

2 The Institution of Structural Engineers. *Structural engineers declare climate and biodiversity emergency.* Available at: www.istructe.org/structural-engineers-declare [Accessed: May 2023].

3 Health and Safety Executive. *The Construction (Design and Management) Regulations 2015.* Available at: www.hse.gov.uk/construction/cdm/2015/index.htm. [Accessed: May 2023].

4 TRICS Consortium. *Trip Rate Information Computer System (TRICS).* Available at: www.trics.org/system.html [Accessed: May 2023].

5 HM Government. *Structure: Approved Document A. A3: Disproportionate collapse*. Available at: https://assets.publishing.service.gov.uk/government/uploads/system/uploads/attachment_data/file/429060/BR_PDF_AD_A_2013.pdf [Accessed: May 2023].

6 The Institution of Structural Engineers. *Structural robustness and disproportionate collapse in buildings* (2nd edition). London: IStructE Ltd, 2023. *In production.*

7 HM Government. *Localism Act 2011.* Available at: www.legislation.gov.uk/ukpga/2011/20/contents/enacted [Accessed: May 2023].

8 HM Government. *National Planning Policy Framework.* 2021. Available at: www.gov.uk/government/publications/national-planning-policy-framework--2 [Accessed: May 2023].

9 HM Government. *Community Infrastructure Levy.* 2022. Available at: www.gov.uk/guidance/community-infrastructure-levy [Accessed: May 2023].

10 HM Government. *The Building Regulations 2010.* Available at: www.legislation.gov.uk/uksi/2010/2214/contents/made [Accessed: May 2023].

11 HM Government. *Equality Act 2010.* Available at: www.legislation.gov.uk/ukpga/2010/15/contents [Accessed: May 2023].

12 Gowler P et al. *Circular economy and reuse: guidance for designers.* London: IStructE Ltd, 2023. *In production.*

13 British Standards Institution. *BS 8500-2:2015+A2:2019. Concrete. Complementary British Standard to BS EN 206 - Specification for constituent materials and concrete.* London: BSI, 2019.

14 British Standards Institution. *BS EN 12464-1:2021. Light and lighting. Lighting of work places – Indoor work places.* London: BSI, 2021.

15 British Standards Institution. *BS 5489-1:2020. Design of road lighting – Lighting of roads and public amenity areas. Code of practice.* London: BSI, 2020.

16 European Parking Association. *European Standard Parking Award Off-Street (ESPA Off-Street).* Available at: www.europeanparking.eu/en/awards/espa-off-street/ [Accessed: May 2023].

17 Safer Parking Scheme/British Parking Association. *Park Mark® safer parking.* Available at: https://parkmark.co.uk/ [Accessed: May 2023].

18 HM Government. *The Traffic Signs Regulations and General Directions 2016.* Available at: www.legislation.gov.uk/uksi/2016/362/contents/made [Accessed: May 2023].

19 British Standards Institution. *BS EN ISO 7010:2020+A6:2023. Graphical symbols: Safety colours and safety signs — Registered safety signs.* London: BSI, 2023.

20 Chrest AP. et al. *Parking Structures: Planning, Design, Construction, Maintenance and Repair* (3rd edition). London; Springer, 2012.

21 RIBA. *RIBA Plan of Work*. Available at: www.architecture.com/knowledge-and-resources/resources-landing-page/riba-plan-of-work [Accessed: May 2023].

22 The Institution of Structural Engineers. *The Structural Plan of Work 2020.* Available at: www.istructe.org/resources/guidance/the-structural-plan-of-work/ [Accessed: May 2023].

23 Designing Buildings: the construction wiki. *Inclusive design.* Available at: www.designingbuildings.co.uk/wiki/Inclusive_design#:~:text=The%20British%20Standards%20Institute%20(2005,special%20adaptation%20or%20specialised%20design.%22 [Accessed: May 2023].

24 HM Government. *Access to and use of buildings: Approved Document M.* Available at: www.gov.uk/government/publications/access-to-and-use-of-buildings-approved-document-m [Accessed: May 2023].

25 British Standards Institution. *BS 8300-2:2018. Design of an accessible and inclusive built environment – Buildings. Code of practice*. London: BSI, 2018.

26 HM Government. *Disability, accessibility and Blve Badge statistics: 2021 to 2022.* Available at: www.gov.uk/blue-badge-scheme-information-council [Accessed: May 2023].
27 Department for Transport. *Disability, accessibility and Blue Badge statistics: 2021 to 2022.* Available at: www.gov.uk/government/statistics/disability-accessibility-and-blue-badge-statistics-2021-to-2022 [Accessed: May 2023].
28 British Standards Institution. *BS 8300-1:2018. Design of an accessible and inclusive built environment – External environment. Code of practice.* London: BSI, 2018.
29 Hill J, Rhodes G and Vollar S. *Car Park Designers' Handbook.* London: ICE Publishing, 2013.
30 The Chartered Institution of Building Services Engineers. *Guide D: Transportation systems in buildings (2020).* London: CIBSE, 2020.
31 British Standards Institution. *BS 9999:2017. Fire safety in the design, management and use of buildings. Code of practice.* London: BSI, 2017.
32 Automobiledimension.com. 2022. Available at: www.automobiledimension.com/car-comparison.php [Accessed: May 2023].
33 HM Government. *Conservation of fuel and power: Approved Document L. Vol.2: Buildings other than dwellings.* Available at: https://assets.publishing.service.gov.uk/government/uploads/system/uploads/attachment_data/file/1099628/ADL2.pdf [Accessed: May 2023].
34 Greater London Authority. *Energy Planning Guidance.* Available at: www.london.gov.uk/programmes-strategies/planning/planning-applications-and-decisions/pre-planning-application-meeting-service/energy-planning-guidance [Accessed: May 2023].
35 CROSS-UK. *CROSS Safety Alert: Fire in multi-storey car parks.* Available at: www.cross-safety.org/uk/safety-information/cross-safety-alert/fire-multi-storey-car-parks [Accessed: May 2023].
36 HM Government. *The Building (Scotland) Regulations 2004.* Available at: www.legislation.gov.uk/ssi/2004/406/contents/made [Accessed: May 2023].
37 HM Government. *The Building Regulations (Northern Ireland) 2012.* Available at: www.legislation.gov.uk/nisr/2012/192/contents/made [Accessed: May 2023].
38 HM Government. *The Regulatory Reform (Fire Safety) Order 2005.* Available at: www.legislation.gov.uk/uksi/2005/1541/contents/made [Accessed: May 2023].
39 HM Government. *BD2552: Fire spread in car parks.* 2010. Available at: https://webarchive.nationalarchives.gov.uk/ukgwa/20120919132719/http://www.communities.gov.uk/documents/planningandbuilding/pdf/1795610.pdf [Accessed: May 2023].
40 Driving Mobility. *ShopMobility UK.* Available at: www.shopmobilityuk.org/ [Accessed: May 2023].
41 Intergovernmental Panel on Climate Change (IPCC). *Special report: Global Warming of 1.5°C.* Available at: www.ipcc.ch/sr15/ [Accessed: May 2023].
42 Intergovernmental Science-Policy Platform on Biodiversity and Ecosystem Services (IPBES). *Global Assessment Report on Biodiversity and Ecosystem Services.* Available at: https://ipbes.net/global-assessment [Accessed: May 2023].
43 HM Government. *25 Year Environment Plan.* Available at: www.gov.uk/government/publications/25-year-environment-plan [Accessed: May 2023].
44 HM Government. *Environment Act 2021.* Available at: www.legislation.gov.uk/ukpga/2021/30/contents/enacted [Accessed: May 2023].
45 The Institution of Structural Engineers. *The hierarchy of net zero design.* Available at: www.istructe.org/resources/blog/the-hierarchy-of-net-zero-design/ [Accessed: May 2023].
46 Burridge J. *How to specify lower carbon concrete.* Available at: www.istructe.org/resources/guidance/how-to-specify-lower-carbon-concrete/ [Accessed: May 2023].
47 Astle P. How can we reduce the embodied carbon of structural concrete? *The Structural Engineer.* 2021:99(2); 24–25. Available at: https://doi.org/10.56330/NWSR6863 [Accessed: May 2023].
48 Swann W. Developing a low-carbon, circular economy for steel. *The Structural Engineer.* 2021:99(4); 18–19. Available at: https://doi.org/10.56330/DPLL5365 [Accessed: May 2023].
49 The Institution of Structural Engineers. *Making your steel specification more sustainable.* Available at: www.istructe.org/resources/guidance/making-your-steel-specification-more-sustainable/#:~:text=You%20can%20consider%20adding%20a,(Gold%20or%20Silver)%20mandatory. [Accessed: May 2023].
50 Gibbons O P and Orr J J *et al. How to calculate embodied carbon* (2nd edition). London: IStructE Ltd, 2022.
51 RICS. *Whole life carbon assessment for the built environment.* Available at: www.rics.org/profession-standards/rics-standards-and-guidance/sector-standards/building-surveying-standards/whole-life-carbon-assessment-for-the-built-environment [Accessed: May 2023].

52 British Standards Institution. *BS EN 15978:2011. Sustainability of construction works. Assessment of environmental performance of buildings. Calculation method.* London: BSI, 2011.
53 Greater London Authority. *Circular Economy Statement Guidance.* Available at: www.london.gov.uk/programmes-strategies/planning/implementing-london-plan/london-plan-guidance/circular-economy-statement-guidance [Accessed: May 2023].
54 Drew C, Fernandez Nova K and Fanning K. *CTBUH Research Paper: The Environmental Impact of Tall vs Small: A Comparative Study.* Council on Tall Buildings and Urban Habitat. 2015. Available at: https://global.ctbuh.org/resources/papers/download/2354-the-environmental-impact-of-tall-vs-small-a-comparative-study.pdf [Accessed: May 2023].
55 MPA The Concrete Centre. 2020. *Specifying Sustainable Concrete.* Available at: www.concretecentre.com/News/2020/Specifying-Sustainable-Concrete.aspx [Accessed: May 2023].
56 HM Government. *Outcome and response to ending the sale of new petrol, diesel and hybrid cars and vans.* Available at: www.gov.uk/government/consultations/consulting-on-ending-the-sale-of-new-petrol-diesel-and-hybrid-cars-and-vans/outcome/ending-the-sale-of-new-petrol-diesel-and-hybrid-cars-and-vans-government-response [Accessed: May 2023].
57 CIRIA. 2015. *The SuDS Manual (C753).* Available at: www.ciria.org/ItemDetail?iProductCode=C753& [Accessed: May 2023].
58 Building Research Establishment. *BREEAM Infrastructure.* Available at: https://bregroup.com/products/ceequal/ [Accessed: May 2023].
59 Building Research Establishment. *Roads and car parks.* Available at: https://bregroup.com/category/case-studies/ceequal-case-studies/ceequal-case-studies-roads-and-car-parks/ [Accessed: May 2023].
60 International Living Future Institute. *Living Building Challenge.* Available at: https://living-future.org/lbc/ [Accessed: May 2023].
61 Building Research Establishment. *Post-Occupancy Evaluation (POE).* Available at: www.bre.co.uk/page.jsp?id=1793 [Accessed: May 2023].
62 BSRIA. *Soft Landings.* Available at: www.bsria.com/uk/consultancy/project-improvement/soft-landings/ [Accessed: May 2023].
63 Arnold W. A short guide to carbon offsetting. *The Structural Engineer.* 2021:99(7); 16–17. Available at: https://doi.org/10.56330/LAUS8964 [Accessed: May 2023].
64 HM Government. *Town and Country Planning Act 1990.* Available at: www.legislation.gov.uk/ukpga/1990/8/contents [Accessed: May 2023].
65 HM Government. *The Building Regulations 2010. The Merged Approved Documents.* Available at: https://assets.publishing.service.gov.uk/government/uploads/system/uploads/attachment_data/file/1082748/Merged_Approved_Documents__Jun2022_.pdf [Accessed: May 2023].
66 HM Government. *Fire safety: Approved Document B.* Available at: www.gov.uk/government/publications/fire-safety-approved-document-b [Accessed: May 2023].
67 HM Government. *Occupiers Liability Act 1984.* Available at: www.legislation.gov.uk/ukpga/1984/3 [Accessed: May 2023].
68 HM Government. *Health and Safety at Work etc. Act 1974.* Available at: www.legislation.gov.uk/ukpga/1974/37/contents [Accessed: May 2023].
69 Disabled Motoring UK. *Disabled Parking Accreditation.* Available at: www.dpaccreditation.org.uk/ [Accessed: May 2023].
70 Society of Automotive Engineers. *Taxonomy and Definitions for Terms Related to Driving Automation Systems for On-Road Motor Vehicles*. Available at: www.sae.org/standards/content/j3016_202104/ [Accessed: May 2023].
71 Via Verde. Available at: www.viaverde.pt/particulares [Accessed: May 2023].
72 HM Government. *Regulation (EU) 2019/2144 of the European Parliament and of the Council.* Available at: www.legislation.gov.uk/eur/2019/2144# [Accessed: May 2023].
73 The Chartered Institution of Building Services Engineers. *Guide M: Maintenance engineering and management (2014).* London: CIBSE, 2014.
74 British Standards Institution. *BS EN 1991-1-1:2002. Eurocode 1. Actions on structures —General actions — Densities, self-weight, imposed loads for buildings.* London: BSI, 2002.
75 British Standards Institution. *NA to BS EN 1991-1-1:2002. UK National Annex to Eurocode 1. Actions on structures — General actions. Densities, self-weight, imposed loads for buildings.* London: BSI, 2019.
76 British Standards Institution. *BS EN 1991-1-4:2005+A1:2010. Eurocode 1. Actions on structures — General actions — Wind actions.* London: BSI, 2005.
77 Willford M R. and Allsop A C. *Design guide for wind loads on unclad framed building structures during construction.* Watford: BRE, 1990.

78 British Standards Institution. *BS EN 1991-1-7:2006+A1:2014. Eurocode 1. Actions on structures — General actions — Accidental actions*. London: BSI, 2006.
79 The Institution of Structural Engineers. *Introduction to structural fire engineering.* London: IStructE Ltd, 2020.
80 British Standards Institution. *BS 6472-1:2008. Guide to evaluation of human exposure to vibration in buildings — Vibration sources other than blasting.* London: BSI, 2008.
81 The Concrete Society. *TR22: Non-structural cracks in concrete* (4th edition). Camberley: The Concrete Society, 2010.
82 The Concrete Society. *TR44: The relevance of cracking in concrete to corrosion of reinforcement* (2nd edition). Camberley: The Concrete Society, 2015.
83 The Concrete Society. *TR67: Movement, restraint and cracking in concrete structures.* Camberley: The Concrete Society, 2008.
84 British Standards Institution. *BS EN 1991-1-5:2003. Eurocode 1. Actions on structures — General actions — Thermal actions.* London: BSI, 2004.
85 The Institution of Structural Engineers. *Manual for the design of building structures to Eurocode 1 & Basis of structural design* (2nd edition). London: IStructE Ltd, 2021.
86 British Standards Institution. *BS EN 1992-1-1:2004+A1:2014. Eurocode 2: Design of concrete structures — General rules and rules for buildings.* London: BSI, 2004.
87 CIRIA. *Control of cracking caused by restrained deformation in concrete (C766).* CIRIA: London, 2018.
88 British Standards Institution. *BS 6093:2006+A1:2013. Design of joints and jointing in building construction. Guide.* London: BSI, 2006.
89 HM Government. *Design Manual for Roads and Bridges — CD 350: The design of highway structures.* Birmingham: National Highways, 2020.
90 MPA The Concrete Centre. *TCC/03/34: Multi-storey Concrete Car Parks.* London: MPA The Concrete Centre, 2012.
91 British Standards Institution. *BS EN 1997-1:2004+A1:2013. Eurocode 7. Geotechnical design — General rules.* London: BSI, 2004.
92 British Standards Institution. *BS 8004:2015+A1:2020. Code of practice for foundations.* London: BSI, 2020.
93 Skempton AW and MacDonald DH. The allowable settlements of buildings. *Proceedings of the Institution of Civil Engineers.* 1956:5(6); 727–768. Available at: https://doi.org/10.1680/ipeds.1956.12202 [Accessed: May 2023].
94 Thorburn S. The real behaviour of structures —soil-structure interaction. *The Structural Engineer.* 1985:63(2); 37–38. Available at: www.istructe.org/journal/volumes/volume-63-(published-in-1985)/issue-2/the-real-behaviour-of-structures-soil-structure-in/ [Accessed: May 2023].
95 Tomlinson MJ. *Foundation Design and Construction* (7th edition). London; Pearson, 2001.
96 Terzaghi K, Peck RB and Mesri G. *Soil Mechanics in Engineering Practice* (3rd edition). Oxford; John Wiley & Sons, Inc., 1996.
97 Field C. Lean yet resilient — designing for the future. *The Structural Engineer.* 2020:98(8); 16–17. Available at: https://doi.org/10.56330/GEZN8281 [Accessed: May 2023].
98 University of Cambridge/University of Bath. *MEICON: Minimising Energy In Construction. Survey of Structural Engineering Practice.* Available at: https://static1.squarespace.com/static/58f72c9a1b631bc0c1e1b84c/t/5b85fb334fa51a1080348b19/1535507448054/MEICON+Report+Online.pdf [Accessed: May 2023].
99 Watson N. Lean design: 10 things to do now. *The Structural Engineer.* 2020:98(8); 13–14. Available at: https://doi.org/10.56330/CDJA1258 [Accessed: May 2023].
100 Building Research Establishment. *Design for Deconstruction — helping construction unlock the benefits of the Circular Economy.* Available at: https://bregroup.com/buzz/design-for-deconstruction-helping-construction-unlock-the-benefits-of-the-circular-economy/#:~:text=Design%20for%20Deconstruction%20(DfD)%20looks,end%20of%20a%20building's%20life [Accessed: May 2023].
101 British Standards Institution. *BS EN 752:2017. Drain and sewer systems outside buildings. Sewer system management.* London: BSI, 2022.
102 British Standards Institution. *BS EN 12056-1:2000. Gravity drainage systems inside buildings — General and performance requirements.* London: BSI, 2000.
103 British Standards Institution. *BS EN 12056-3:2000. Gravity drainage systems inside buildings — Roof drainage, layout and calculation.* London: BSI, 2000.
104 British Standards Institution. *BS 6213:2000+A1:2010. Selection of construction sealants. Guide.* London: BSI, 2000.
105 British Standards Institution. *BS EN 206:2013+A2:2021. Concrete. Specification, performance, production and conformity.* London: BSI, 2021.

106 British Standards Institution. *BS 8500-1:2015+A2:2019. Concrete. Complementary British Standard to BS EN 206 — Method of specifying and guidance for the specifier.* London: BSI, 2020.
107 British Standards Institution. *BS EN ISO 12944-4:2017. Paints and varnishes. Corrosion protection of steel structures by protective paint systems — Types of surface and surface preparation.* London: BSI, 2018.
108 BRE Construction Division. *Concrete in aggressive ground.* Watford: BRE, 2017.
109 The Institution of Structural Engineers. *Structural effects of alkali-silica reaction.* London: IStructE Ltd, 1992.
110 The Thaumasite Expert Group. The structural Implications of the thaumasite form of sulfate attack. *The Structural Engineer.* 1999:77(4); 10–11. Available at: www.istructe.org/journal/volumes/volume-77-(published-in-1999)/issue-4/the-structural-implications-of-the-thaumasite-form/ [Accessed: May 2023].
111 Building Research Establishment. *Corrosion of steel in concrete: durability of reinforced concrete structures (DG 444-1).* Watford: BRE, 2000.
112 Building Research Establishment. *Corrosion of steel in concrete: investigation and assessment (DG 444-2).* Watford: BRE, 2000.
113 Building Research Establishment. *Corrosion of steel in concrete: protection and remediation (DG 444-3).* Watford: BRE, 2000.
114 Henderson NA, Johnson RA and Wood JGM. Enhancing the Whole Life Structural Performance of Multi-Storey Car Parks. 2002. Available at: https://assets.publishing.service.gov.uk/government/uploads/system/uploads/attachment_data/file/429311/enhancing_the_whole_life_structural_performance_of_multi-storey_car_parks.pdf [Accessed: May 2023].
115 British Standards Institution. *BS 8204-2:2003+A2:2011. Screeds, bases and in situ floorings — Concrete wearing surfaces. Code of practice*. London: BSI, 2003.
116 The Concrete Society. *Concrete practice: guidance on the practical aspects of concreting (GCG8) (2nd edition).* Camberley: The Concrete Society, 2016.
117 British Standards Institution. *BS EN 14889-2:2006. Fibres for concrete — Polymer fibres. Definitions, specifications and conformity.* London: BSI, 2006.
118 British Standards Institution. *BS EN 14889-1:2006. Fibres for concrete — Steel fibres. Definitions, specifications and conformity.* London: BSI, 2006.
119 The Concrete Society. *TR66: External in-situ concrete paving.* Camberley: The Concrete Society, 2007.
120 The Concrete Society. *TR73: Cathodic protection of steel in concrete, including model specification.* Camberley: The Concrete Society, 2011.
121 The Concrete Society. *Concrete Advice No.7: Galvanised steel reinforcement.* Camberley: The Concrete Society, 2021.
122 British Standards Institution. *BS ISO 14654:1999. Epoxy-coated steel for the reinforcement of concrete.* London: BSI, 2000.
123 The Concrete Society. *TR72: Durable post-tensioned concrete structures*. Camberley: The Concrete Society, 2010.
124 British Standards Institution. *BS EN ISO 12696:2022. Cathodic protection of steel in concrete.* London: BSI, 2022.
125 The Institution of Structural Engineers. *Structural use of fibre polymer composites.* London: IStructE Ltd, 2023.
126 The Concrete Society. *TR75: Composite concrete slabs using steel decking.* Camberley: The Concrete Society, 2016.
127 MPA The Concrete Centre. *Concrete Basements.* London: MPA The Concrete Centre, 2012.
128 British Standards Institution. *BS EN 1504-5:2013. Products and systems for the protection and repair of concrete structures. Definitions, requirements, quality control and evaluation of conformity — Concrete injection.* London: BSI, 2013.
129 British Standards Institution. *Multi-part Document: BS EN 1504. Products and systems for the protection and repair of concrete structures. Definitions, requirements, quality control and evaluation of conformity.* London: BSI, 2005–2017.
130 British Standards Institution. *BS EN 1504-10:2017. Products and systems for the protection and repair of concrete structures. Definitions, requirements, quality control and evaluation of conformity — Site application of products and systems and quality control of the works.* London: BSI, 2017.
131 British Standards Institution. *BS EN 13670:2009. Execution of concrete structures.* London: BSI, 2010.
132 CONSTRUCT. *National Structural Concrete Specification for Building Construction (NSCS)* (4th edition). Available at: https://construct.org.uk/toolkit/national-structural-concrete-specification-for-building-construction-edition-4/ [Accessed: May 2023].
133 British Standards Institution. *BS EN 13892-4:2002. Methods of test for screed materials — Determination of wear resistance-BCA.* London: BSI, 2002.
134 British Standards Institution. *BS 6150:2019. Painting of buildings. Code of practice.* London: BSI, 2019.

135 British Standards Institution. *BS EN 1504-2:2004. Products and systems for the protection and repair of concrete structures. Definitions, requirements, quality control and evaluation of conformity — Surface protection systems for concrete.* London: BSI, 2004.
136 British Standards Institution. *BS 8204-5:2004+A1:2011. Screeds, bases and in situ floorings — Mastic asphalt underlays and wearing surfaces. Code of practice.* London: BSI, 2004.
137 Health and Safety Executive. *Control of Substances Hazardous to Health (COSHH).* Available at: www.hse.gov.uk/coshh/ [Accessed: May 2023].
138 British Standards Institution. *BS EN 13501-1:2018. Fire classification of construction products and building elements — Classification using data from reaction to fire tests.* London: BSI, 2019.
139 European Organisation for Technical Assessment. Available at: www.eota.eu/about-us [Accessed: May 2023].
140 The Concrete Society. *TR50: Guide to surface treatments for protection and enhancement of concrete.* Camberley:
The Concrete Society, 1997.
141 The Institution of Structural Engineers. *Temporary demountable structures: Guidance on procurement, design and use*
(4th edition). London: IStructE Ltd, 2017.
142 Health and Safety Executive. *Assessing the slip resistance of flooring.* Available at: www.hse.gov.uk/pubns/geis2.htm [Accessed: May 2023].
143 British Standards Institution. *Multi-part Document: PD 6688 — Recommendations for the design of structures to BS EN 1991-1.* London: BSI, 2007–2015.
144 HM Government. *Protection from falling, collision and impact: Approved Document K.* Available at: www.gov.uk/government/publications/protection-from-falling-collision-and-impact-approved-document-k [Accessed: May 2023].
145 HM Government. *Edge protection in multi-storey car parks: Design specification and compliance testing.* Available at: www.gov.uk/government/publications/edge-protection-in-multi-storey-car-parks-review-of-design-standards [Accessed: May 2023].
146 HM Government. *Edge protection in multi-storey car parks — Assessment method for installed restraint systems in multi-storey car parks.* Available at: www.gov.uk/government/publications/edge-protection-in-multi-storey-car-parks [Accessed: May 2023].
147 British Standards Institution. *BS 6180:2011. Barriers in and about buildings. Code of practice.* London: BSI, 2011.
148 The Chartered Institution of Building Services Engineers. *Guide A: Environmental design (2015, updated 2021).* London: CIBSE, 2015.
149 HM Government. *Ventilation: Approved Document F.*
Available at: www.gov.uk/government/publications/ventilation-approved-document-f [Accessed: May 2023].
150 British Standards Institution. *BS 7346-7:2013. Components for smoke and heat control systems — Code of practice on functional recommendations and calculation methods for smoke and heat control systems for covered car parks.* London: BSI, 2013.
151 The Chartered Institution of Building Services Engineers. *SLL Lighting Handbook (2018).* London: CIBSE, 2018.
152 British Standards Institution. *Multi-part Document: BS EN 13201 — Road lighting.* London: BSI, 2015–2016.
153 British Standards Institution. *BS EN 1838:2013. Lighting applications. Emergency lighting.* London: BSI, 2013.
154 British Standards Institution. *BS 5266-1:2016. Emergency lighting — Code of practice for the emergency lighting of premises.* London: BSI, 2016.
155 Building Services Research and Information Association. *Life Safety and Firefighting Power Supplies (BG 70/2021).* Bracknell: BISRIA, 2021.
156 British Standards Institution. *BS 5839-1:2017. Fire detection and fire alarm systems for buildings — Code of practice for design, installation, commissioning and maintenance of systems in non-domestic premises.* London: BSI, 2018.
157 British Standards Institution. *BS 8519:2020. Selection and installation of fire-resistant power and control cable systems for life safety, fire-fighting and other critical applications. Code of practice.* London: BSI, 2020.
158 British Standards Institution. *BS EN 62305-1:2011. Protection against lightning — General principles.* London: BSI, 2011.
159 British Standards Institution. *BS 7671:2018+A2:2022. Requirements for Electrical Installations. IET Wiring Regulations.* London: BSI, 2022.
160 British Standards Institution. *BS EN 12845:2015+A1:2019. Fixed firefighting systems. Automatic sprinkler systems. Design, installation and maintenance.* London: BSI, 2020.

161 Strakosch G R and Caporale R S. *The vertical transportation handbook* (4th edition). John Wiley & Sons, Inc.: Oxford, 2010.
162 British Standards Institution. *BS EN 81-22:2021. Safety rules for the construction and installation of lifts. Lifts for the transport of persons and goods — Passenger and goods passenger lifts with inclined travel path.* London: BSI, 2021.
163 British Standards Institution. *BS 5655-6:2011. Lifts and service lifts — Code of practice for the selection, installation and location of new lifts.* London: BSI, 2011.
164 British Standards Institution. *BS EN 12101-6:2022. Smoke and heat control systems — Specification for pressure differential systems. Kits*. London: BSI, 2022.
165 British Standards Institution. *BS EN 1363-1:2020. Fire resistance tests — General requirements.* London: BSI, 2020.
166 The Institution of Structural Engineers. *Guide to the advanced fire safety engineering of structures.* London: IStructE Ltd, 2007.
167 HM Government. *Fire performance of green roofs and walls*. Available at: www.gov.uk/government/publications/fire-performance-of-green-roofs-and-walls [Accessed: May 2023].
168 Chitty R. *External fire spread: building separation and boundary distances (BR 187 2nd edition).* Watford: BRE, 2014.
169 British Standards Institution. *PAS 1899:2022. Electric vehicles — Accessible EV charging points. Specification.* London: BSI, 2022.
170 British Standards Institution. *PAS 6463:2022. Design for the mind. Neurodiversity and the built environment. Guide.* London: BSI, 2022.
171 Samaritans. *Latest suicide data*. Available at: www.samaritans.org/about-samaritans/research-policy/suicide-facts-and-figures/latest-suicide-data/ [Accessed: May 2023].
172 Police Crime Prevention Initiatives. *Secured by Design (SBD).* Available at: www.securedbydesign.com/ [Accessed: May 2023].
173 HM Government. *Preventing suicides in public places: A practice resource.* Available at: www.gov.uk/government/publications/suicide-prevention-suicides-in-public-places [Accessed: May 2023].
174 British Standards Institution. *BS 8000-0:2014. Workmanship on construction sites — Introduction and general principles.* London: BSI, 2014.
175 British Standards Institution. *BS 8000-2.1:1990. Workmanship on building sites. Code of practice for concrete work — Mixing and transporting concrete.* London: BSI, 1990.
176 British Standards Institution. *BS 8000-2.2:1990. Workmanship on building sites. Code of practice for concrete work — Sitework with in situ and precast concrete.* London: BSI, 1990.
177 The Concrete Society. *TR62: Self-compacting concrete: A review.* Camberley: The Concrete Society, 2005.
178 Cement Admixtures Association. *Admixture Technical Sheet — ATS 5: Air-entraining admixtures.* Available at: www.admixtures.org.uk/download/ats-5-air-entraining/ [Accessed: May 2023].
179 Health and Safety Executive. *Electrical standards and approved codes of practice.* Available at: www.hse.gov.uk/electricity/standards.htm [Accessed: May 2023].
180 Health and Safety Executive. *Pipers Row Car Park, Wolverhampton — Quantitative study of the causes of the partial collapse on 20 March 1997.* Available at: www.hse.gov.uk/research/misc/index.htm [Accessed: May 2023].
181 Whapples C and McKibbins L. (Eds.) *Recommendations for the Inspection, Maintenance and Management of Car Park Structures* (2nd edition). London: ICE Publishing, 2018.
182 British Parking Association. *Parking Practice Note 17: Asset Management and Maintenance of Parking Structure.* Haywards Heath: BPA, 2005.
183 British Parking Association. *Parking Practice Note 6: Life-Care Plan for Parking Structures.* Haywards Heath: BPA, 2006.
184 National Trust for Historic Preservation. *The Greenest Building: Quantifying the Environmental Value of Building Reuse.* Available at: https://forum.savingplaces.org/viewdocument/the-greenest-building-quantifying [Accessed: May 2023].
185 Manchester Evening News. *Multi-storey car park to be converted into 116 homes including rows of rooftop townhouses.* Available at: www.manchestereveningnews.co.uk/news/greater-manchester-news/multi-storey-car-park-converted-24244493 [Accessed: May 2023].